gets you through

# A-LEVEL
# MATHS
## YEAR 2
## IN A WEEK

7 DAYS

**ROSIE BENTON**
**SHARON FAULKNER**

# CONTENTS

# Indices and Surds

You should be able to convert easily between different forms of numbers and expressions. This skill underpins many of the more complex topics at A-level, so a thorough understanding is needed. Do not rush, as many students drop marks by losing focus on the relatively simple parts of a more complex problem.

## Laws of Indices

At A-level, recognising, interpreting and manipulating indices is important and it is an assumed skill within many other topics. It is also an area where marks are often lost because students either rush it or haven't understood it fully in the first place.

| Laws | Examples |
|------|----------|
| $x^a \times x^b = x^{a+b}$ | $2^2 \times 2^3$ <br> $= 2 \times 2 \times 2 \times 2 \times 2$ <br> $= 2^5$ |
| $x^a \div x^b = x^{a-b}$ | $2^3 \div 2^5 = \frac{\cancel{2} \times \cancel{2} \times \cancel{2}}{\cancel{2} \times \cancel{2} \times \cancel{2} \times 2 \times 2}$ <br> $= \frac{1}{2^2}$ <br> $= 2^{-2}$ |
| $x^{-a} = \frac{1}{x^a}$ | $2^{-3} = \frac{1}{2^3} = \frac{1}{8}$ |
| $(x^a)^b = x^{ab}$ | $(2^2)^3$ <br> $= 2^2 \times 2^2 \times 2^2$ <br> $= 2 \times 2 \times 2 \times 2 \times 2 \times 2$ <br> $= 2^6$ <br> $= 2^{2 \times 3}$ |
| $x^{\frac{a}{b}} = \sqrt[b]{x^a} = \left(\sqrt[b]{x}\right)^a$ | $25^{\frac{3}{2}} = \left(\sqrt[2]{25}\right)^3 = (5)^3 = 125$ |

## Applying the Laws and Manipulating Indices

The rules above can be used to interpret and to simplify an expression, whether algebraic or numerical.

### Example

Express $18^2 \times 3^{\frac{2}{3}} \div 4$ in the form $3^p$, where $p$ is a rational number to be found.

If unsure about how to start, it is worth considering using prime factorisation of numbers as this should show where common factors are. In this case, the question suggests that the final form should have 3 as the base, so finding a way to write the other bases in terms of 3 could also be a good starting point.

$$18^2 \times 3^{\frac{2}{3}} \div 4 = (2 \times 3^2)^2 \times 3^{\frac{2}{3}} \div 2^2$$
$$= 2^2 \times 3^4 \times 3^{\frac{2}{3}} \div 2^2$$
$$= 3^{4 + \frac{2}{3}}$$
$$= 3^{\frac{14}{3}}$$

Note: Indices are left as improper fractions rather than converted to mixed numbers.

## Evaluating Expressions in the Form $x^{-\frac{a}{b}}$

It is best to think about these types of question as three distinct steps:

⚫ **Step 1:** Is there a negative? If yes, use the reciprocal of $x$.
⚫ **Step 2:** Is there a denominator? If yes, consider taking the associated root next.
⚫ **Step 3:** Raise the whole of your base to the power.

### Example

Evaluate $\left(\frac{9}{16}\right)^{-\left(\frac{3}{4}\right)}$, giving your answer in the form $\frac{a}{\sqrt{b}}$.

The negative in the power can be 'used' to flip the fraction upside-down:

$$\left(\frac{9}{16}\right)^{-\left(\frac{3}{4}\right)} = \left(\frac{16}{9}\right)^{\left(\frac{3}{4}\right)}$$

The fourth root can be 'invoked' to find the fourth root of both the numerator and the denominator:

$$= \left(\frac{\sqrt[4]{16}}{\sqrt[4]{9}}\right)^3 = \left(\frac{2}{\sqrt{3}}\right)^3$$

Both numerator and denominator are cubed:

$$= \frac{2^3}{\sqrt{3}^3} = \frac{8}{\sqrt{3}^3} = \frac{8}{\sqrt{27}}$$

All exams allow a calculator, so use it to check your answer. However, be aware of questions that prompt for clear working or exact answers. Understanding how the rules work and what the different elements in an index mean will help to answer questions that are not simple to key into a calculator. A calculator will always give the simplest form, so exam questions will include algebraic terms in the powers, have answers that aren't fully simplified (questions may specify the form of the answer, as in the previous example), or ask to see the individual steps taken in the working. If a question says 'show that', detailed mathematical reasoning must be shown and the result from a calculator will not gain the marks.

## Surds

A surd is a representation of a number using a root sign (most often the square root but it could be any root). Surds are used to accurately represent irrational numbers (i.e. numbers that can't be written accurately as a fraction). If a question asks for the exact answer, it will involve surds or $\pi$ or e. There are more irrational numbers but none as significant as these.

Surds are just numbers and, as such, they obey all the arithmetic and algebraic rules. They should be written clearly with any rational multiplier before the surd; this helps to make it clear what is within the root and what isn't.

There are certain mathematical conventions to adhere to when dealing with surds. As ever, the aim is to simplify a number or expression as much as possible. Mathematically, the answer to the previous example $\left(\frac{8}{\sqrt{27}}\right)$ is not yet complete. Key it into a calculator and it might suggest the number is $\frac{8\sqrt{3}}{9}$. There are two key steps in this manipulation:

### 1. Simplify a Surd

If the number under a square root is a square number, then square root it.

**Example**

$\sqrt{9} = 3$

If the number under a square root has a factor that is a square number, it can be simplified.

**Example**

$\sqrt{18} = \sqrt{9} \times \sqrt{2} = 3\sqrt{2}$

The same would apply to cube, fourth, etc. roots.

**Example**

$\sqrt[5]{96} = \sqrt[5]{32} \times \sqrt[5]{3} = 2\sqrt[5]{3}$

Familiarity with square and cube numbers is helpful here.

### 2. Rationalise the Denominator

Mathematically the convention is that, for a final answer, all surds should be represented in the numerator, with none in the denominator.

*Rationalising Simple Denominators (the case where the denominator is a single surd, i.e. $\sqrt{a}$)*

In the case where the original number is $\frac{q}{a\sqrt{b}}$, where both $a$ and $b$ are rational, it can be multiplied by $\frac{\sqrt{b}}{\sqrt{b}} = 1$. Multiplying by 1 does not change the value of the original number but can be used to change the form in which that number is written.

$$\frac{q}{a\sqrt{b}} \times \frac{\sqrt{b}}{\sqrt{b}} = \frac{q \times \sqrt{b}}{a\sqrt{b} \times \sqrt{b}} = \frac{q\sqrt{b}}{ab}$$

**Example**

$\frac{4\sqrt{3}}{\sqrt{2}} = \frac{4\sqrt{3}}{\sqrt{2}} \times \frac{\sqrt{2}}{\sqrt{2}} = \frac{4\sqrt{3 \times 2}}{2} = \frac{4\sqrt{6}}{2} = 2\sqrt{6}$

*Rationalising More Complex Denominators*

When the denominator contains the sum or the difference of two numbers, at least one of which is a surd, the difference of two squares is used:

$$x^2 - y^2 = (x + y)(x - y)$$

In effect, both squares become rational, and there are no $x$ or $y$ terms.

If the denominator of a fraction is $(a + b)$, where $a$, $b$ or both are irrational, the denominator needs to be multiplied by $(a - b)$, so the numerator must be multiplied by the same.

$$\frac{q}{a\sqrt{b}+c\sqrt{d}} \times \frac{a\sqrt{b}-c\sqrt{d}}{a\sqrt{b}-c\sqrt{d}} = \frac{q(a\sqrt{b}-c\sqrt{d})}{a^2b-c^2d}$$

**Example**

Rationalise the denominator $\dfrac{4}{\sqrt{2}-3\sqrt{5}}$

$$\frac{4}{\sqrt{2}-3\sqrt{5}} \times \frac{\sqrt{2}+3\sqrt{5}}{\sqrt{2}+3\sqrt{5}} = \frac{4(\sqrt{2}+3\sqrt{5})}{2-9\times5}$$

$$= \frac{4\sqrt{2}+12\sqrt{5}}{2-45}$$

$$= -\frac{4\sqrt{2}+12\sqrt{5}}{43}$$

### Links to Other Concepts

- Ranges and domains of functions – trigonometry
- Problem solving in context (maximisation problems)
- Quadratics, solving and manipulating – discriminant
- Problems in mechanics and statistics
- Solving equations
- Logs and exponentials
- Integration/differentiation
- Transformation of graphs
- Coordinate geometry
- Pythagoras' Theorem

## SUMMARY

- Positive integer indices represent repeated multiplications of the base number.
- $x^{-a} = \dfrac{1}{x^a}$
- $x^a \times x^b = x^{a+b}$
- $x^a \div x^b = x^{a-b}$
- $(x^a)^b = x^{ab}$
- $x^{\frac{a}{b}} = \sqrt[b]{x^a} = \left(\sqrt[b]{x}\right)^a$
- Surds represent a set of irrational numbers. They are often needed to write a number 'exactly' (i.e. without rounding).
- $\sqrt{a} \times \sqrt{a} = a$, so the square root of a number is the number that when multiplied by itself gives the value of the number under the root sign.
- Surds can be simplified by extracting square factors from inside a square root (or cube numbers from inside a cube root).
- Conventionally, surds should be represented in the numerator of a fraction. To rationalise the denominator, multiply the denominator and the numerator by the same value. If the denominator is $a + \sqrt{c}$ then the multiplier is $a - \sqrt{c}$.
- Converting from index form can help to understand numbers better. Converting into index form means numbers can be combined and then differentiated or integrated more easily.

## QUICK TEST

1. Express each of the following in index form, i.e. $q^a$.

   a) $\sqrt[4]{q^3}$

   b) $\dfrac{1}{q^2}$

   c) $\dfrac{1}{q^2} \times \sqrt[4]{q^3}$

   d) $\left(\dfrac{1}{q^4}\right)^3$

2. Which is the correct first step when differentiating the function $y = \dfrac{5x}{\sqrt[3]{x^2}} - \dfrac{1}{\sqrt[3]{27x}}$?

   A $\quad y = 5x \div x^{-\frac{2}{3}} - 3x^{\frac{1}{3}} \quad \rightarrow \quad 2x^{-\frac{19}{24}}$

   B $\quad y = 5x \times x^{-\frac{2}{3}} - 27x^{\frac{1}{3}} \quad \rightarrow \quad 5x^{\frac{5}{3}} - 3x^{\frac{1}{3}}$

   C $\quad y = 5x \div x^{\frac{2}{3}} - \dfrac{x^{-\frac{1}{3}}}{3} \quad \rightarrow \quad 5x^{\frac{1}{3}} - \dfrac{x^{-\frac{1}{3}}}{3}$

   D $\quad y = 5x \times x^{-\frac{2}{3}} - 3x^{-3} \quad \rightarrow \quad 5x^{\frac{1}{3}} + 3x^3$

3. Express $\left(\sqrt[2]{x^5}\right)^{-\frac{2}{5}}$ without using index form.

4. Rationalise the denominator of the following:

   a) $\dfrac{2}{\sqrt{3}}$

   b) $\dfrac{\sqrt{3}}{\sqrt{5} - 2\sqrt{3}}$

5. Express each of the following in the form $k \times 2^b$, where $k$ is an expression.

   a) $\dfrac{1}{\sqrt{8}}$

   b) $\dfrac{\sqrt{36}}{\sqrt[3]{16a^3}}$

6. Find the scale factor for the enlargement of these two similar rectangles:

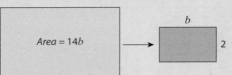

## PRACTICE QUESTIONS

1. a) Rationalise the denominator of the expression $\dfrac{3\sqrt{a}}{3\sqrt{b} - \sqrt{a}}$, where $a$ and $b$ are rational constants. **[2 marks]**

   b) Hence or otherwise, find the values of $a$ and $b$ where $\dfrac{3\sqrt{a}}{\sqrt{a} - 3\sqrt{b}} = \dfrac{18\sqrt{5} + 15}{31}$ **[3 marks]**

2. Given that $\cos(30) = \dfrac{\sqrt{3}}{2}$, write $(2^a \cos(30))^{-\frac{4}{a}}$ in the form $\sqrt[a]{\dfrac{1}{p \times q^{a-1}}}$, where $p$ and $q$ are values to be found. **[5 marks]**

3. A family of quadratic curves is given by the equation $px^2 + 3px + p = 0$.

   a) Use the quadratic formula $\left(\dfrac{-b \pm \sqrt{b^2 - 4ac}}{2a}\right)$ to find expressions for the roots as a single rational function. **[2 marks]**

   b) Hence find the fully rationalised values of $\dfrac{1}{x}$ given that $x$ satisfies the equation $6x^2 + 18x = -6$. Show your working clearly. **[4 marks]**

# Polynomials, Quadratics and Simultaneous Equations

## Simplifying Polynomials

'Like' terms have the same power of $x$ and can be added or subtracted to simplify the polynomial. Write the polynomial in decreasing powers of $x$.

Polynomial expressions can be represented as a product of factors, but the factors will also be polynomials or will be binomials. You must be able to manipulate a polynomial to express it both in full form but also as a product of factors. The full form makes it easy to differentiate or integrate. In factorised form it is easy to find roots of the polynomial, and the factors are clear.

## Expanding Brackets

The factors expressed within brackets will expand out to show the full polynomial.

> **Example**
> Expand $(2x + 3)(x - 1)(3x + 2)$.
> $= (2x^2 + 3x - 2x - 3)(3x + 2)$
> $= (2x^2 + x - 3)(3x + 2)$
> $= 6x^3 + 4x^2 + 3x^2 + 2x - 9x - 6$
> $= 6x^3 + 7x^2 - 7x - 6$

## Factorising Polynomials

Methods for factorising polynomials include **comparing coefficients**, **polynomial division**, and **inspection**.

> **Don't forget**
> Look out for simple factorisations where each term in an expression or equation contains a common factor, e.g. $6x^4 + 4x^2 - 10x = 2x(3x^3 + 2x - 5)$.

## Factor Theorem

The factor theorem is used to identify a linear (binomial) factor of a polynomial.

If $f(x) = (x + b)g(x)$, then substituting in $x = -b$ would give $f(-b) = (-b + b)g(x) = 0 \times g(x) = 0$.

Factor theorem states that if $f(a) = 0$, then $(x - a)$ will be a factor of the expression $f(x)$.

> **Example**
> Show that $(x - 1)$ is a factor of the expression $3x^3 + 2x - 5$.
>
> $f(1) = 3 \times 1^3 + 2 \times 1 - 5 = 0$
>
> $\therefore (x - 1)$ is a factor.

## Simple Algebraic Division

Having identified a factor $(f(x) = (x + b)g(x))$, algebraic division is one method for finding the function $g(x)$.

> **Example**
>
> $$5x \div x$$
> $$3x^2 \div x$$
> $$3x^3 \div x$$
> $$3x^2 + 3x + 5$$
> $$x - 1 \overline{\smash{\big)}\ 3x^3 + 0x^2 + 2x - 5}$$
> $3x^2 \times (x - 1) \rightarrow \quad -\underline{(3x^3 - 3x^2)}$
> $$0 + 3x^2 + 2x$$
> $3x \times (x - 1) \rightarrow \quad -\underline{(3x^2 - 3x)}$
> $$5x - 5$$
> $5 \times (x - 1) \rightarrow \quad -\underline{(5x - 5)}$
> $$0$$

Including any terms with coefficient 0 is important. Take extra care with negatives.

## Comparing Coefficients

An alternative method to algebraic division is comparing coefficients.

> **Example**
> $3x^3 + 2x - 5 = (x - 1)(ax^2 + bx + c)$
> $\qquad = ax^3 + bx^2 - ax^2 + cx - bx - c$
> $\qquad = ax^3 + (b - a)x^2 + (c - b)x - c$
>
> For these two expressions to be equal, the coefficient of each power of $x$ must be equal.

Equating $x^3$ terms:     $a = 3$
Equating $x^2$ terms:     $b - a = 0 \rightarrow b = 3$
Equating $x$ terms:     $c - b = 2 \rightarrow c = 5$
Equating $x^0$ terms:     $-c = -5 \rightarrow c = 5$

$f(x) = (x - 1)(3x^2 + 3x + 5)$

At this stage the expression is written as two factors, a linear (or binomial) and a quadratic.

## Representing Quadratics

Quadratics can be presented in a variety of forms:

- **Fully expanded form**, e.g. $y = ax^2 + bx + c$
  If sketching the graph, it gives the $y$-intercept ($c$) and shows if the graph is positive or negative ($a$).
- **Factorised form**, e.g. $y = (dx + e)(fx + g)$
  This is useful as it gives the $x$-intercepts – the 'roots' of the equation – in this case $-\frac{e}{d}$ and $-\frac{g}{f}$.
- **Completed square form**, e.g. $y = (x + h)^2 + k$
  This form can be used to find the maximum or minimum point. The variable is contained within a square so the lowest value the bracket can take is 0. Therefore the lowest value $y$ can have is $k$; this happens when $x = -h$, i.e. the turning point of the quadratic is $(-h, k)$.

## Solving Quadratics

To solve a quadratic, get all the terms on one side, simplified in an equation equal to zero. Quadratics can have either two distinct real roots, a repeated real root or no real roots. If there are real roots to the equation, you can find them using one of three methods:

### 1. Factorisation
Further factorisation means that a quadratic can be split into two brackets, either of which could be zero to get the desired result. It is always worth looking for factors as it can be the simplest and quickest way to get to the result. If factors aren't obvious, use the quadratic formula or completing the square.

### 2. Completing the Square
Completing the square allows all the terms involving an unknown to be represented in one place. It can be used with the generalised quadratic equation $(ax^2 + bx + c)$ to find the quadratic formula, which can be applied to any quadratic.

### 3. The Quadratic Formula

$$x = \frac{-b \pm \sqrt{b^2 - 4ac}}{2a}$$

### The Discriminant

The **discriminant** allows a relatively simple calculation to determine if there are roots to a quadratic equation.

- If $b^2 - 4ac > 0$ there are two real distinct roots, represented by the $x$-intercepts on the graph.

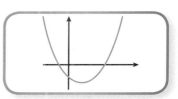

- If $b^2 - 4ac = 0$ there is a repeated real root.

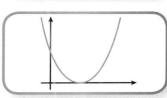

- If $b^2 - 4ac < 0$ there are no real roots to the quadratic equation.

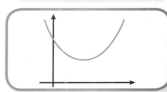

The discriminant can be used to decide if there are roots to a quadratic, or to help find missing terms in a quadratic given the number of real roots.

**Example**
The equation $px^2 + 3x + 2$ has a repeated root. Find the value of $p$.

$$3^2 - 4 \times 2 \times p = 0$$
$$8p = 9$$
$$p = \frac{9}{8}$$

## Hidden Quadratics

An unknown value could be represented by any letter or even by an expression. Sometimes it is easier to define an expression with a letter in order to reveal and solve the quadratic. This is okay as long as the letter is defined clearly and at the point of getting to an answer it is reinterpreted in terms of the question.

## Example

$3\sin^4 x + \sin^2 x = 2\sin^2 x + 1 - 3\sin^4 x$. Find all the possible values of $\sin x$.

At first glance this doesn't necessarily leap out as being a quadratic equation. As there is a $\sin^2 x$ and a $\sin^4 x$, it is possible to make a substitution which will help to reveal the quadratic:

Let $a = \sin^2 x$

$3a^2 + a = 2a + 1 - 3a^2$

For any equation, simplify by collecting like terms.

$6a^2 - a - 1 = 0$

Having revealed the quadratic, solve as such:

$(2a - 1)(3a + 1) = 0$

$(2a - 1) = 0 \rightarrow a = \frac{1}{2}$

$(3a + 1) = 0 \rightarrow a = -\frac{1}{3}$

This is solved for $a$, but since $a = \sin^2 x$, the question is not yet answered. Substitute $\sin^2 x$ back into the results:

$a = -\frac{1}{3} \rightarrow \sin^2 x = -\frac{1}{3}$

There are no valid real results as $\sin^2 x \neq -\frac{1}{3}$

$a = \frac{1}{2} \rightarrow \sin^2 x = \frac{1}{2} \rightarrow \sin x = \frac{\pm\sqrt{2}}{2}$

Any result where there are two variants of the unknown can be treated as a quadratic. For example, $4x^{\frac{4}{3}} + 2x^{\frac{2}{3}} - 7 = 0$ can be treated as a quadratic. A possible substitution is to let $a = x^{\frac{2}{3}}$, which would give a simplified equation of $4a^2 + 2a - 7 = 0$. Substitutions aren't necessary but can help to make working clearer.

## Solving Simultaneous Equations

To solve equations with more than one unknown, you need as many independent equations as there are unknowns. So for two unknowns, you need two independent equations involving the same variables.

An equation with two simple variables (no index greater than 1) has an infinite number of possible solution pairs. Graphically this is represented by a straight line.

If an independent equation (that is, not just a rearrangement of the same equation) is available, it defines a second relationship between the variables. The intersection of the two graphs gives solutions that fit both equations. If both equations are linear, there will be a single point of intersection – just one pair of values that makes both equations true. If one of the equations describes a quadratic curve, there may be two distinct solutions, one repeat solution, or no solutions.

There are two main methods used to solve simultaneous equations algebraically:

### 1. Elimination

Elimination works well if there is the same magnitude coefficient for one of the unknowns. In this case the two equations can be added or subtracted in order to 'eliminate' that variable, meaning there is a single equation to work with containing only one unknown. Elimination can also be used when there is a simple multiplication that can be made to one or both equations to give coefficients of equal magnitude.

### 2. Substitution

Substitution works for all simultaneous equations. By rearranging one equation to find an expression for one variable in terms of the other, you can substitute this expression into the other equation. This is the only option when one of the equations is a polynomial or other non-linear graph. If both equations are linear, it doesn't matter which one is rearranged. Otherwise, it is generally easier to rearrange the linear equation and substitute into the quadratic.

## Example

$4x - 2y - 3 = 0$ and $x + y^2 + 2y - 16 = 0$

Find the values of $x$ and $y$, giving each to 3 significant figures.

Rearrange the linear equation to isolate one of the unknown terms:

$2y = 4x - 3 \qquad \rightarrow \qquad y = 2x - \frac{3}{2}$

Substitute this into the other equation:

$x + \left(2x - \frac{3}{2}\right)^2 + 2\left(2x - \frac{3}{2}\right) - 16 = 0$

Expand and simplify to form a quadratic and solve:

$x + 4x^2 - 6x + \frac{9}{4} + 4x - 3 - 16 = 0 \rightarrow$
$$4x^2 - x - 16.75 = 0$$

$a = 4, b = -1, c = -16.75$

$x = \dfrac{1 \pm \sqrt{1 - 4 \times 4 \times (-16.75)}}{8} = \dfrac{1 \pm \sqrt{269}}{8}$

Substitute the fully accurate value (either in surd form or from calculator memory) into either equation; generally the linear will be simpler:

$$4 \times \frac{1+\sqrt{269}}{8} - 2y - 3 = 0 \rightarrow$$

$$y = \frac{-5+\sqrt{269}}{4} = 2.85 \text{ (3 s.f.)}, x = 2.18 \text{ (3 s.f.)}$$

$$4 \times \frac{1-\sqrt{269}}{8} - 2y - 3 = 0 \rightarrow$$

$$y = \frac{-5-\sqrt{269}}{4} = -5.35 \text{ (3 s.f.)}, x = -1.93 \text{ (3 s.f.)}$$

## SUMMARY

- Polynomials have multiple terms and are written with the highest power of $x$ (or other variable/unknown) first.

- To expand a set of brackets, take a pair at a time and multiply each term in the first bracket by every term in the second bracket.

- To factorise a polynomial, factor theorem can be used to identify a factor: $f(a) = 0$ then $(x - a)$ will be a factor of the expression $f(x)$.

- Having found a factor, algebraic division, comparing coefficients and inspection are all valid methods of taking the factorising step.

- Quadratics are generally factorised by inspection.

- Quadratics can be solved using factorisation, completing the square, or the quadratic formula, $x = \frac{-b \pm \sqrt{b^2 - 4ac}}{2a}$

- Using the discriminant:

  $b^2 - 4ac > 0$, there are two real distinct roots

  $b^2 - 4ac = 0$, there is a repeated real root

  $b^2 - 4ac < 0$, there are no real roots to the quadratic equation.

- The key methods for solving simultaneous equations are elimination and substitution. Graphically the solutions are represented by the points where two lines cross.

### Links to Other Concepts
- Solving problems   ● Equation of a circle
- Finding roots and solutions
- Inequalities   ● Trigonometric equations
- Mechanics and laws of motion   ● Statistics

## QUICK TEST

1. $x^2 - 4x + y^2 + 2y - 20 = 0$ and $x - y - 4 = 0$. Find the solutions for $x$ and $y$.

2. Fully factorise the polynomial $f(x) = 2x^3 - x^2 - 13x - 6$ given that $f(-2) = 0$.

## PRACTICE QUESTIONS

1. A straight line has an equation $y = \frac{1}{2}x - 2$ and a curve has the equation $y = x^2 - 2x - 3$.

   a) Use a sketch to show that both intercepts of the two graphs have negative $y$-coordinates. **[5 marks]**

   b) Show that the $x$-coordinates satisfy the equation $2x^2 - 5x - 2 = 0$. **[2 marks]**

2. $6x^3 + 20x^2 + 17x + 2 = 0$

   a) Use factor theorem to show that a solution is $x = -2$ and hence express the cubic in the form $(x + 2)(px^2 + qx + r) = 0$. **[4 marks]**

   b) Use the method of completing the square to find the exact solutions to the equation $6x^2 + 8x + 1 = 0$, giving your answers as single, simplified fractions. Show your working clearly. **[5 marks]**

   c) Hence or otherwise, find the roots to the equation $10a^{1.2} + 8a^{0.6} - 2 = 4a^{1.2} - 3$, giving answers to 3 significant figures. **[5 marks]**

# Inequalities and Partial Fractions

## Simple Inequalities

An inequality describes a condition where there are a range of possible solutions. If there is one variable involved in an inequality, it can be solved like an equation but with some special considerations:

- When rearranging the inequality, consider moving the terms past the fixed inequality symbol.

- The only time the inequality symbol reverses is when dividing or multiplying by a negative.

- Common sense needs to be applied; consider the context to make sense of a result.

- Substituting in a value as a check will highlight if there are any mistakes.

- Quadratics and simultaneous equations could be involved in inequalities. Diagrams can help!

**Example**

**Method 1:**
$$4+3x<29-2x$$
$$4+3x+2x<29$$
$$5x<29-4$$
$$x<5$$

**Method 2: (demonstrating the effect of a negative):**
$$4+3x<29-2x$$
$$4-29<-2x-3x$$
$$-25<-5x$$
$$25>5x$$
$$5>x$$

## Using Set Notation to Express Inequalities

Set notation can be used to describe a set of numbers:

- $x \in \{x : x < 5\}$ means $x$ belongs to (is an element of) a set of numbers that are less than 5.

- $x \in \{x : x < 5 \text{ or } x \geq 8\}$ means $x$ belongs to a set of numbers that are less than 5, or greater than or equal to 8.

- $x \in \{x : x < 5\} \cup \{x \geq 8\}$ means that $x$ is an element of the union of the two sets, so $x$ can be in either.

- $x \in (-\infty, 5) \cup [8, \infty)$ also refers to a union of two sets; the numbers in the brackets represent the limits of the set – a curved bracket is not included in the set but a square bracket is.

## Quadratic Inequalities

Quadratic inequalities can be treated as an equation. Solve them using the methods from the last topic in order to find the key points (where the inequality changes from being true to false) and then put the inequality signs back at the end.

**Example**
$$x^2 + x \geq 6$$
$$x^2 + x - 6 \geq 0$$
$$(x+3)(x-2) \geq 0$$

$y = x^2 + x - 6$

$x^2 + x - 6 \geq 0$

So $x \leq -3$, $x \geq 2$

If the equation had been $x^2 + x \leq 6$ then the final answer would be $-3 \leq x \leq 2$. This is written together as it is a single section on the graph. The original answer is two separate areas so is written as two separate inequalities.

## Simultaneous Inequalities and Diagrams

You could be asked for the conditions under which two inequalities are both true at the same time.

**Example**
$$4+3x<29-2x \text{ and } x^2+x \geq 6$$

The solutions to each of these individually were found above. To consider them together, a diagram on a number line can be used. A line, drawn parallel to the number line either above or below, is used to represent the values of $x$ that would make the individual inequalities true. A small empty circle is used on the end to represent < and >, but a coloured-in circle is used for $\leq$ and $\geq$.

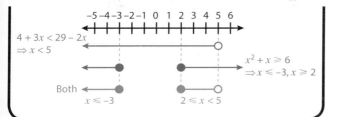

$4 + 3x < 29 - 2x$
$\Rightarrow x < 5$

$x^2 + x \geqslant 6$
$\Rightarrow x \leqslant -3, x \geqslant 2$

Both

$x \leqslant -3$    $2 \leqslant x < 5$

# Inequalities with Two Variables

You can represent inequalities involving two variables on a graph. Consider the inequality as an equation and use a grid of values or the form $y = mx + c$ to plot the line. The points on each line represent the solutions to the equation and shading is used to indicate the side of the line which gives solutions to the inequality:

- If the line is dotted, it means the points on the line are not included ($<$ or $>$).

- If the line is solid, it means the points on the line are possible solutions ($\leqslant$ or $\geqslant$).

## Example

Shade the region $y + 2x - 3 < 0$ on a graph.

Start by considering the line $y + 2x - 3 = 0$, which can be rewritten in the form $y = -2x + 3$, a line passing through 3 on the $y$-axis and with a gradient of $-2$. A table of results could also be used to draw the line:

| $x$ | $-1$ | $0$ | $1$ | $2$ |
|---|---|---|---|---|
| $y$ | $5$ | $3$ | $1$ | $-1$ |

A dotted line is used to represent this line as the solutions to the equation are not valid for the inequality ($<$). The area below the line is shaded.

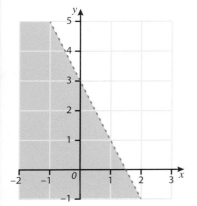

Quick check – select any point in the shaded area, e.g. (1, 0) and substitute into the inequality: $0 + (2 \times 1) - 3 < 0$? Yes, this is true.

Some questions ask for a region defined by a number of inequalities. Draw each as before but with only a small amount of shading, or short arrows, to indicate which side of the line is the required region until all lines are in place. Then shade the area required.

## Example

Shade the region defined by $y \leqslant x$, $x < 2$, $y \geqslant \frac{1}{2}$

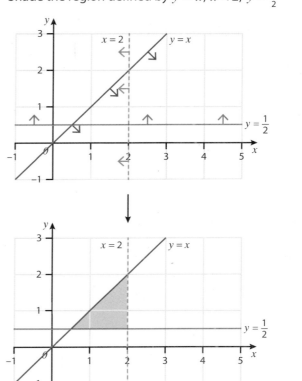

# Algebraic Fractions

Algebraic fractions can be simplified by finding common factors between the numerator and the denominator. See pages 8–11 in relation to manipulating polynomials.

## Example

Simplify the algebraic fraction $\dfrac{(2x^2 + 3x + 1)(x - 3)}{(x^2 - 6x + 9)(x + 1)}$

Use clues from the question to help wherever possible. In this case, the denominator is likely to have $(x - 3)$ as a factor, or the numerator $(x + 1)$, perhaps even both.

$$\frac{(2x + 1)(x + 1)(x - 3)}{(x - 3)(x - 3)(x + 1)} = \frac{2x + 1}{x - 3}$$

Some questions could ask for the final form to include a remainder. Use remainder theorem and a method of algebraic division.

**Remainder theorem** says that when dividing by $(ax + b)$, the remainder will be $f\left(-\frac{b}{a}\right)$.

**Factor theorem** is a specific case of the remainder theorem, when $f\left(-\frac{b}{a}\right) = 0$; there is no remainder so $(ax + b)$ is a factor.

---

**Example**

Express the algebraic fraction $\frac{4x^3 + 4x^2 - 3x + 3}{2x + 3}$ in the form $ax^2 + bx + \frac{c}{2x + 3}$.

**Method 1: Long division**

$$
\begin{array}{r}
2x^2 - x \phantom{+0000} \\
2x + 3 \overline{\big)\; 4x^3 + 4x^2 - 3x + 3} \\
-(4x^3 + 6x^2) \phantom{00000} \\
\hline
-2x^2 - 3x \phantom{000} \\
-(-2x^2 - 3x) \phantom{00} \\
\hline
0 \;+\; 3 \;(\text{remainder})
\end{array}
$$

$$\Rightarrow\; 2x^2 - x + \frac{3}{2x + 3}$$

**Method 2: Comparing coefficients**

$$4x^3 + 4x^2 - 3x + 3 = (2x + 3)\left(ax^2 + bx + \frac{c}{2x + 3}\right)$$

$$4x^3 + 4x^2 - 3x + 3 = 2ax^3 + 3ax^2 + 2bx^2 + 3bx + c$$

| | |
|---|---|
| $x^3$ coefficients: | $4 = 2a \rightarrow a = 2$ |
| $x^2$ coefficients: | $3a + 2b = 4 \rightarrow b = -1$ |
| $x$ coefficients: | $3b = -3 \rightarrow b = -1$ ✓ |
| $x^0$ coefficients: | $c = 3$ |

$$\frac{4x^3 + 4x^2 - 3x + 3}{2x + 3} = 2x^2 - x + \frac{3}{2x + 3}$$

---

## Partial Fractions

Partial fractions are used to express a complicated algebraic fraction as the sum of a number of simpler algebraic fractions. This produces a form that can be integrated and differentiated much more easily. If the numerator and the denominator are polynomials and the degree (the highest power in the polynomial) of

the numerator is more than that of the denominator, you will need to use algebraic division first.

The first step in partial fractions is to **simplify** where possible. Look for common factors that will cancel.

---

**Example**

Express $\frac{(22x^2 + 32x + 9)}{(3x^3 + 8x^2 + 7x + 2)}$ as partial fractions.

First, fully factorise the denominator $g(x)$. By inspection $g(-1) = 0$ so $(x + 1)$ is a factor.

$$\frac{f(x)}{g(x)} = \frac{(22x^2 + 32x + 9)}{(x + 1)(3x^2 + 5x + 2)} = \frac{(22x^2 + 32x + 9)}{(x + 1)(3x + 2)(x + 1)}$$

$$= \frac{(22x^2 + 32x + 9)}{(3x + 2)(x + 1)^2}$$

⬤ If the denominator is of the form $(ax + b)(cx + d)(ex + f)$, the fraction can be split into partial fractions in the form
$$\frac{A}{ax + b} + \frac{B}{cx + d} + \frac{C}{ex + f}$$

⬤ If the denominator is of the form $(ax + b)(cx + d)^2$, the fraction can be split into partial fractions in the form
$$\frac{A}{ax + b} + \frac{B}{cx + d} + \frac{C}{(cx + d)^2}$$

$$\frac{(22x^2 + 32x + 9)}{(3x + 2)(x + 1)^2} = \frac{A}{3x + 2} + \frac{B}{x + 1} + \frac{C}{(x + 1)^2}$$

**Method 1: Substituting values of $x$ that make the brackets $= 0$**

$$22x^2 + 32x + 9 = A(x + 1)^2 + B(3x + 2)(x + 1) + C(3x + 2)$$

When $x = -1$: $\quad 22 - 32 + 9 = 0A + 0B + C(-1)$

$$\rightarrow C = \mathbf{1}$$

When $x = -\frac{2}{3}$: $\quad 22 \times \frac{4}{9} + 32 \times -\frac{2}{3} + 9$

$$= A\left(\frac{1}{3}\right)^2 + B(0)\left(\frac{1}{3}\right) + C(0)$$

$$-\frac{23}{9} = \frac{A}{9} \rightarrow A = \mathbf{-23}$$

To find $B$, consider any value for $x$ along with the values found for $A$ and $C$. Choose a simple number.

When $x = 0$:

$$22 \times 0 + 32 \times 0 + 9 = (-23)(1)^2 + B(2)(1) + 1(2)$$

$$9 = -23 + 2B + 2$$

$$2B = 30 \rightarrow B = \mathbf{15}$$

## Method 2: Comparing coefficients

Multiply both sides by the denominator:

$$22x^2 + 32x + 9 = A(x+1)^2 + B(3x+2)(x+1) + C(3x+2)$$

$$= Ax^2 + 2Ax + A + 3Bx^2 + 5Bx + 2B + 3Cx + 2C$$

Equating coefficients gives:

$$22 = A + 3B$$
$$32 = 2A + 5B + 3C$$
$$9 = A + 2B + 2C$$

This gives simultaneous equations to solve:

$$A = 22 - 3B$$
$$32 = 2(22 - 3B) + 5B + 3C$$
$$32 = 44 - B + 3C$$
$$\mathbf{B = 3C + 12}$$
$$A = 22 - 3(3C + 12)$$
$$A = \mathbf{-14 - 9C}$$
$$9 = (-14 - 9C) + 2(3C + 12) + 2C$$
$$9 = 10 - C$$
$$\mathbf{C = 1, A = -23, B = 15}$$

Give the final answer in the required form:

$$\frac{f(x)}{g(x)} = \frac{-23}{3x+2} + \frac{15}{x+1} + \frac{1}{(x+1)^2}$$

### Links to Other Concepts
- Comparing coefficients of polynomials
- Trigonometry ● Set notation and statistics
- Binomial expansion ● Calculus
- Algebraic manipulation ● Factor theorem

1. Solve and represent these inequalities on a number line.
   a) $4 > 3 - x$  b) $2x^2 - x \leqslant x^2 - 3x + 8$
   c) $4 > 3 - x$ and $2x^2 - x \leqslant x^2 - 3x + 8$
2. Which form will the partial fractions of the following expression take? $\dfrac{3x-2}{(x-3)(5x+1)(x-3)}$

## PRACTICE QUESTIONS

1. Find the values of $A$, $B$ and $C$ given that:
   $12x - 2 = A(x-1)(2x+1) + B(x-1)(3x+5) + C(2x+1)(3x+5)$ **[4 marks]**

2. On graph paper, shade the region represented by $y \geqslant 2x^2 - 11x + 5$ and $y + x < 5$. **[4 marks]**

3. a) Show that $(3x + 1)$ is a factor of $6x^2 - 13x - 5$. **[2 marks]**

   b) Simplify the fraction $\dfrac{34x^2 - 68x}{12x^3 - 26x^2 - 10x}$ **[3 marks]**

   c) Hence or otherwise, express $\dfrac{34x^2 - 68x}{12x^3 - 26x^2 - 10x}$ as partial fractions. **[6 marks]**

4. The simultaneous inequalities $ax - b < 0$ and $(x+2)\left(x - \dfrac{b}{a+5}\right) \geqslant 0$, where $a$ and $b$ are positive integers, are represented as follows:

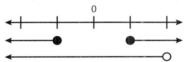

The solution to the simultaneous inequalities is $x \leqslant -2$ and $2 \leqslant x < 7$. Find the values of $a$ and $b$. **[7 marks]**

## SUMMARY

- Set notation, diagrams and symbols can all be used to represent an inequality.

- Partial fractions are a way to represent complex algebraic fractions with simplified denominators, making the expressions more straightforward for applying calculus.

  – If the denominator is of the form $(ax + b)(cx + d)(ex + f)$, the fraction can be split into partial fractions in the form $\dfrac{A}{ax+b} + \dfrac{B}{cx+d} + \dfrac{C}{ex+f}$

  – If the denominator is of the form $(ax + b)(cx + d)^2$, the fraction can be split into partial fractions in the form $\dfrac{A}{ax+b} + \dfrac{B}{cx+d} + \dfrac{C}{(cx+d)^2}$

- Factor theorem and algebraic division can help to establish factors in the denominator.

- The remainder theorem says that when dividing by $(ax + b)$, the remainder will be $f\left(-\dfrac{b}{a}\right)$. If $f\left(-\dfrac{b}{a}\right) = 0$, then $(ax + b)$ is a factor.

# Functions, Graphs and Transformations

Functions describe a relationship between an input variable (often $x \in X$) and an output variable (often $y \in Y$). A mapping connects one element from set $X$ (the **domain**) to one element in set $Y$ (the **range**).

A **one-to-one** function has exactly one element from $X$ that maps onto exactly one element from $Y$.

**Example**

$f(x) = 2x$, $x \in \mathbb{R}$

The domain is all real numbers and the range is also all real numbers $y \in \mathbb{R}$.

A **many-to-one** function has a number of elements from $X$ that map onto an element from $Y$.

**Example**

$f(x) = x^2$, $x \in \mathbb{R}$

The domain is all real numbers and the range is $\{y : y \geqslant 0\}$

If a single element from the domain maps onto more than one element from the range, then it is **not a function**.

**Example**

$f(x) = \sqrt{x}$, $x \in \mathbb{R}$

For some functions, the domain is limited.

**Example**

$f(x) = x^{-\frac{1}{2}} + 3$

In this case the domain is $\{x : x > 0\}$ and the range is $\{y : y > 3\}$.

## Inverse Functions

If a function is one-to-one, you can find its inverse function. A many-to-one function would become one-to-many and therefore not be a function.

To find an inverse function, $f^{-1}(x)$, of the original function, $f(x)$, rearrange so that the original input (domain) becomes the output (range) and vice-versa. Or put another way, rearrange to make $x$ the subject of the equation, then swap the $x$s for $y$s and vice-versa. Graphically, it is a reflection of the function in the line $y = x$. The domain of the original function becomes the range of the inverse function. The range of the original function becomes the domain of the inverse function.

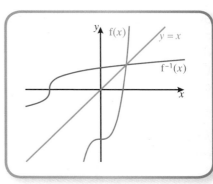

**Example**

Find the inverse function of $y = 3x^3 - 2$, $x \in \mathbb{R}$.

$3x^3 = y + 2$

$x^3 = \dfrac{y+2}{3}$

$x = \sqrt[3]{\dfrac{y+2}{3}}$

The inverse function is $y = \sqrt[3]{\dfrac{x+2}{3}}$, $x \in \mathbb{R}$.

## Composite Functions

A composite function is formed when two, or more, functions are applied. Functions tend to be called $f(x)$, $g(x)$ and $h(x)$ but could be referred to in terms of other letters or notation. A composite function can be described as $fg(x)$, or "f of g of x". Always apply the functions from the closest to the brackets outwards.

**Example**

Find $fg(x)$ given that $f(x) = x^2 - 2$ and $g(x) = 2x - 3$.

$fg(x) = f(g(x))$

$f(2x - 3) = (2x - 3)^2 - 2$

$= 4x^2 - 6x - 6x + 9 - 2$

$= 4x^2 - 12x + 7$

# Drawing Graphs

Graphs are a visual representation of functions. Graphs can help to model the behaviour and predict outcomes. They also show where key elements are, such as **roots**, **asymptotes** and **turning points**.

**Sketches** should include key elements, clearly labelled, and the basic shape. They don't need to be presented on scaled axes and don't waste time on too much detail. **Drawing** is more accurate and may require the plotting of points; generally a scaled axis will be provided.

## Straight-Line Graphs

A **linear function** is represented by a straight line on a graph. The general form is $y = mx + c$. The key features to include on a sketch are:

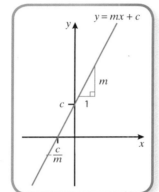

- the $y$-intercept, $c$

- the $x$-intercept, $-\frac{c}{m}$.

## Quadratics

A quadratic has a general form $y = ax^2 + bx + c$. The key features to include on a sketch are:

- the $y$-intercept, $c$

- the $x$-intercepts, also known as 'roots of the equation', which can be found by setting $y = 0$ and then solving the quadratic.

The completed square form is an easy way to find the maximum or the minimum point. As the quadratic curve is symmetrical, you can identify the maximum or the minimum point by locating the midpoint between the $x$-intercepts and substituting the value found into the original equation to find the corresponding $y$-value.

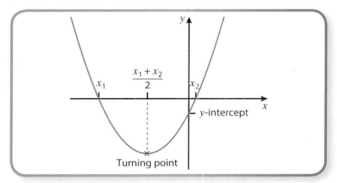

## Other Polynomials

Each additional power in a polynomial adds a turning point. Key features to be included on a sketch are axis-intercepts and turning points. There are three types of turning point; **maxima**, **minima** and **points of inflection**. In some cases, these turning points occur in the same place. For example, $y = x^4$ might look as though it has one minimum point but in fact it has three turning points in the same place. Graphical calculators can help with sketching, but knowing the shapes and possibilities for solutions is useful. See the table below.

| Graph of $y = f(x)$ where: | General shape | | Possible solution / roots? | | |
|---|---|---|---|---|---|
| | $a$ is positive | $a$ is negative | | | |
| $y = ax^2 + bx + c$ (quadratic) | ∪ | ∩ | No roots | Repeat root | Two distinct roots |
| $y = ax^3 + bx^2 + cx + d$ (cubic) | Max. Min. / Point of inflection | | One root | Root / Repeat root | Three distinct roots |

The discriminant can be used with a quadratic to find how many roots there are. The factorised form also gives this information.

### Example

$y = x^2 + 4x + 4$ factorises to $y = (x + 2)(x + 2)$.

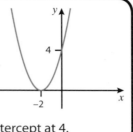

Both brackets give a result of $x = -2$, so there is a repeated root at $-2$. It is a positive quadratic with a $y$-intercept at 4.

The factorised form will also give this information for other polynomials.

### Example

$y = (2x - 3)^2(x + 1)$ would give a curve that is a positive cubic passing through $x = -1$ then dipping down to touch at $x = \frac{3}{2}$ (as this is the 'repeated root').

To find the $y$-intercept, multiply the constants: $-3 \times -3 \times 1 = 9$.

If more detail is needed, for example on which side of the $y$-axis the turning point is, you can use differentiation to locate the turning point precisely. If the exact detail isn't important, substitute a relatively small value of $x$ into the equation on either side:

$(2 \times 0.1 - 3)^2(0.1 + 1) = 8.62$ (3 s.f.) and
$(2 \times -0.1 - 3)^2(-0.1 + 1) = 9.22$ (3 s.f.)

This shows the graph is decreasing as it passes through the $y$-axis.

### Reciprocal Curves

Reciprocal curves have a function of $x$ in the denominator when expressed as $y = \frac{1}{f(x)}$. Any value divided by 0 is considered to be undefined, so **asymptotes** are produced where the $x$-value makes the denominator equal zero.

### Example

$y = \frac{1}{x + 2}$ has a vertical asymptote of $x = -2$.

As the equation can be rearranged to $f(x) = \frac{1}{y}$, the line $y = 0$ also gives an asymptote. Asymptotes are drawn onto the axes with a dotted line, which represents the value that the curve will approach but never touch.

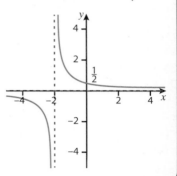

Another key reciprocal graph is $y = \frac{1}{x^2}$. It also has asymptotes and these lie on the axes.

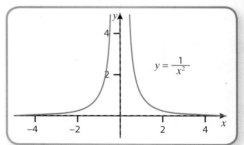

Note: A curve can sometimes cross an asymptote before approaching its value. In the graph below, $y = 0$ is an asymptote as $x$ tends to $\pm \infty$ but $y = 0$ for a value $-1 < x < 0$.

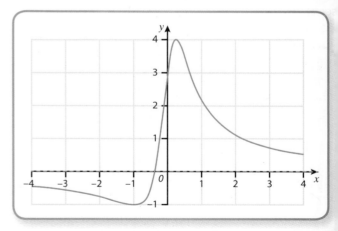

### Proportionality

Two values being directly **proportional** will result in a straight-line graph, as $y \propto x$ gives an equation $y = kx$, where $k$ is a constant:

- If $y$ is proportional to a function of $x$, the graph will be in the shape represented by $f(x)$ but stretched by a factor of $k$.

- If the variables are inversely proportional, a reciprocal graph is created, as $y \propto \frac{1}{x}$ gives an equation $y = \frac{1}{kx}$ or $y = \frac{k}{x}$.

## Exponential Graphs

Graphs of the form $y = a^x$ are exponential:

- If $a > 1$, the curve passes through 1 on the $y$-axis. As $x$ decreases, the curve tends towards the $x$-axis, which forms an asymptote.

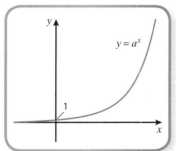

- If $0 < a < 1$, the curve is effectively reflected in the $y$-axis.

## Modulus

The modulus of a function is the magnitude of the function at any given point, but without the 'direction' (positive/negative) – so all values are treated as positive. Vertical line brackets are used to show a modulus function.

To plot a modulus function, $y = |f(x)|$, any negative sections of the original graph (i.e. parts that lie below the $x$-axis) are reflected in the $x$-axis and so become positive.

### Example
The graph shown is $y = |-0.5x + 1|$.

The line plotted is:

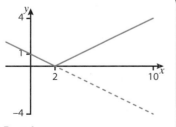

For $x \leqslant 2$     $y = -0.5x + 1$

For $x \geqslant 2$     $y = -(-0.5x + 1) = 0.5x - 1$ (from graph transformation $-f(x)$, which gives a reflection in the $x$-axis)

In questions that ask for simultaneous solutions involving a modulus, try to consider the graphs.

### Example
Find the points of intersection of $y = |-0.5x + 1|$ and $y = |3x - 9|$.

Let $f(x) = -\frac{1}{2}x + 1$ and $g(x) = 3x - 9$.

The sketch shows the plots of the two modulus functions.

From the sketch you can see that there are two possible solutions to the problem. One lies on the section of the line which is the $-g(x)$ and $-f(x)$.

$$\frac{1}{2}x - 1 = -3x + 9$$

$$\frac{7}{2}x = 10$$

$$x = \frac{10}{\frac{7}{2}} = \frac{20}{7}$$

To find the $y$-value, substitute back into one of the equations:

$$y = \frac{1}{2} \times \frac{20}{7} - 1 = \frac{3}{7}$$

For the second solution, use $-f(x)$ and set it equal to $+g(x)$:

$$\frac{1}{2}x - 1 = 3x - 9$$

$$\frac{5}{2}x = 8$$

$$x = \frac{16}{5}$$

Then find the associated $y$-value:

$$y = \frac{1}{2}\left(\frac{16}{5}\right) - 1 = \frac{3}{5}$$

Points of intersection are $\left(\frac{20}{7}, \frac{3}{7}\right)$ and $\left(\frac{16}{5}, \frac{3}{5}\right)$.

## Transformations

Using a general function in terms of $x$, i.e. $y = f(x)$, you can create graphs which have the same traits but have been transformed in some way. Transformations may include **reflections, stretches parallel to the axes** and **translations**. You should be familiar with the transformations $y = f(x) + a$, $y = f(x + a)$, $y = af(x)$, $y = f(ax)$, $y = -f(x)$ and $y = f(-x)$ from Year 1.

With certain graphs it is possible for two different transformations to give the same result:

- If $f(x) = a^x$, then $f(x+b) = a^{x+b} = a^b \times a^x$. This transformation could be described as either a translation or a stretch.
- If $f(x) = mx$, then $f(x+1) = mx + m = f(x) + m$. This could be considered to be either a translation of $\begin{pmatrix} -1 \\ 0 \end{pmatrix}$ or of $\begin{pmatrix} 0 \\ m \end{pmatrix}$.

### Example
Describe the transformation from $y = 3^x$ to $y = 3^{x+2}$ as a stretch.

$$3^{x+2} = 3^2 \times 3^x = 9f(x)$$

The transformation is a stretch of scale factor 9 in the $y$-direction.

## Combining Transformations

Transformations can be applied individually or in combination. Some combinations involve movement or stretching that is at right angles to each other, so it doesn't matter which way round the two transformations are applied.

### Example
The transformation of $f(x)$ onto $g(x)$ is described as $g(x) = 2f(x-1)$. Describe the nature of the transformation.

The 2 outside the original function means a stretch in the $y$-direction with scale factor 2. The $-1$ inside the brackets means a translation of 1 space to the right.

These two transformations would not impact upon each other so can be applied in either order and the resulting graph would still be the same.

If the two transformations are in the same direction (horizontal or vertical), then it becomes important to know in which order they are applied.

### Example
Starting from the graph of $y = |x|$

| $y = |x| - 2$ | $y = |x - 2|$ |
|---|---|
| 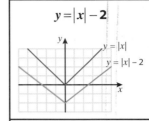 | 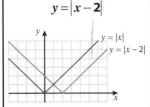 |
| $y = 3|x|$ | $y = |3x|$ |
| 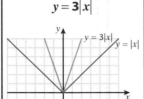 | 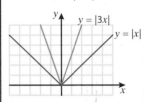 |

| $y = 3|x| - 2$ | $y = |3x - 2|$ |
|---|---|
| Both affect the vertical. Transformations are applied intuitively with BIDMAS. First apply the stretch, then the translation. 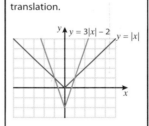 | Both parts of the transformation affect the horizontal. The order is now counter-intuitive, i.e. against BIDMAS. The translation happens first, then the stretch.<br><br>It might be easier to consider this case as the line $y = 3x - 2$ with the negative region reflected. Some functions may not be so obvious though! 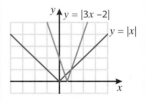 |

## Links to Other Concepts
- Comparing coefficients of polynomials  - Trigonometric functions  - Differentiation and integration
- Simultaneous equations  - Ranges and domains of functions – trigonometry  - Problem solving in context
- Solving and manipulating quadratics – discriminant  - Problems in mechanics and statistics
- Numerical methods, finding roots of equations  - Vector notation  - Manipulation of indices
- Exponentials and logs  - Factor theorem, algebraic division  - Algebraic manipulation

## SUMMARY

- Functions can be one-to-many or one-to-one mappings.
- The inverse function is a reflection in the line $y = x$:
  - The domain of $f(x)$ becomes the range of $f^{-1}(x)$
  - The range of $f(x)$ becomes the domain of $f^{-1}(x)$
  - Only one-to-one mappings have an inverse function.
- Composite functions are where more than one function is applied. The order is important: for $gh(x)$, apply $h(x)$ first, then apply $g(x)$ to the result.
- Sketches are important to model and understand questions.
- Key features that should be included on a graph are intercepts, asymptotes and turning points.
- Polynomials form continuous curves. The roots of the equation can be found by factorisation or other numerical methods. Factor theorem can help to identify a linear factor for a cubic graph.
- If there is a repeat factor, this represents a repeat root on the graph.
- The modulus of a number is a positive number of equal magnitude.
- To sketch $y = |f(x)|$, reflect all sections that lie below the $x$-axis to above the $x$-axis.
- To sketch $ay = f(|x|)$, consider the transformations.
- Transformations can be combined. The order of the transformations is insignificant if they affect the perpendicular directions:
  - If both transformations affect the horizontal direction, the order is counter-intuitive regarding order of operations
  - If both transformations affect the vertical direction, the order is intuitive regarding order of operations.

## QUICK TEST

1. $y = f(x)$ gives the graph

   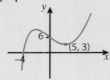

   For both of these, define $y$ in terms of $f(x)$

   a) 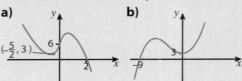 b)

2. The graph (right) shows a polynomial of $y = x^2(x+a)(x+b)$. State the values of $a$ and $b$.

   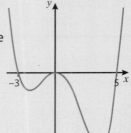

3. Sketch the graph of $y = |2x + 3|$.

4. Sketch the graph of $y = -\frac{1}{2}|x + 2|$.

5. What is the range of the function $y = 3x^2 - 3$, $-2 < x < 4$?

6. $f(x) = 3x + 1$, $g(x) = x^2 - 2$, $x \in \mathbb{R}$

   a) Find $ff(x)$. b) Find $gf(x)$.

   c) Sketch the graph of $y = -|g(x) - 2|$.

   d) Find the function $f^{-1}(x)$.

## PRACTICE QUESTIONS

1. a) Find the solutions to the equations
   $y = |2x + 3|$ and $y = |x - 2|$ **[6 marks]**

   b) Hence solve the inequality
   $|2x + 3| < |x - 2|$ **[2 marks]**

2. The graph of $y = x^2$ is horizontally stretched by scale factor 2 and then translated $\begin{pmatrix} 0 \\ -3 \end{pmatrix}$, resulting in the composite function $fg(x^2)$.

   a) Given that $f(x) = x - 2$, what is the function $g(x)$? **[4 marks]**

   b) What is the $y$-coordinate on the graph of $y = fg(x^2)$ when $x = 4$? **[3 marks]**

# Coordinate Geometry

## Straight Lines

If an equation has $x$ and $y$ variables with index of 1, it is a linear relationship and is represented on a set of axes by a straight line. The equation of a straight line can be written in different forms, the three most common being:

**1.** $y = mx + c$

This form allows the gradient ($m$) and the $y$-intercept ($c$) to be found without any further calculations.

> **Example**
> The equation $y = -3x + 0.2$ has a gradient of $-3$ and a $y$-intercept of 0.2.

**2.** $ax + by + c = 0$

This is the form that means that $a$, $b$ and $c$ can all be integer values. It is often the required form in an exam question. Calculations are required to find the gradient and $y$-intercept:

- ⚫ To find the gradient, use $\frac{-a}{b}$
- ⚫ To find the $y$-intercept, use $\frac{-c}{b}$.

**3.** $y = a, x = b$

These are simplifications of the form $y = mx + c$, but where the coefficient of one of the variables is equal to zero.

> **Example**
> $y = -2$ gives a straight line parallel to the $x$-axis and crossing through $-2$ on the $y$-axis. If considered as part of $y = mx + c$, it would be $y = 0x - 2$, so the gradient is 0 (horizontal) and the $y$-intercept is $-2$.

## Gradients

The gradient of a line is how many spaces it goes up for each space it goes across. A gradient of $\frac{-1}{2}$ means it goes down half a space for every whole space you move to the right. It is the change in $y$ divided by the change in $x$.

A pair of **parallel** lines have the same gradient.

A pair of **perpendicular** lines (i.e. at right angles to each other) have gradients that are **negative reciprocals**. This means the product of their gradients is $-1$.

The gradient function (differentiated function) is another way of finding the gradient of a line. See pages 58–63.

### Finding the Equation of a Straight Line

$y_2 - y_1 = m(x_2 - x_1)$ can be used to find the equation of a straight line if you are given two points, $(x_1, y_1)$ and $(x_2, y_2)$, that the line passes through or if you are given the gradient (or a way in which to find the gradient) and a point the line passes through. This leads towards an equation in the form $ax + by + c = 0$.

To find the gradient between two points, use $m = \frac{y_2 - y_1}{x_2 - x_1}$. It doesn't matter which point is $(x_1, y_1)$ and which is $(x_2, y_2)$ as long as they are used within their pairs consistently.

> **Example**
> A line goes through two points, $P(-2, 7)$ and $Q(4, 3)$. Find the equation of the line in the form $ax + by + c = 0$, where $a$, $b$ and $c$ are integer values to be found.
>
> Find the gradient: $m = \frac{(y_2 - y_1)}{(x_2 - x_1)} = \frac{7 - 3}{-2 - 4}$
> $$= \frac{4}{-6} = \frac{-2}{3}$$
>
> To find the equation, substitute the gradient, a known point and $(x, y)$, a general point on the line, into $y_2 - y_1 = m(x_2 - x_1)$. It is generally simplest to substitute the known point as $(x_1, y_1)$.
>
> $$y - 7 = \frac{-2}{3}(x - -2)$$
>
> Then simplify:
>
> $$3y - 21 = -2x - 4$$
> $$2x + 3y - 17 = 0$$
> $$a = 2, b = 3 \text{ and } c = -17$$

## Length and Midpoint

Pythagoras' Theorem can be used to find the **length** of the line segment, as the two variables ($x$ and $y$) are at right angles.

## Example

Find the length of the line segment between the points $C(-2, 5)$ and $D(3, -1.5)$ correct to 2 decimal places.

A sketch can help to show what to do.

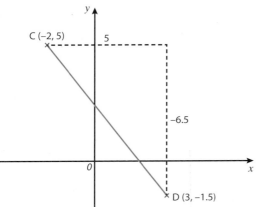

Pythagoras' Theorem states $a^2 + b^2 = c^2$. This question is asking for the length of the hypotenuse so you can use $c = \sqrt{a^2 + b^2}$.

Length $\sqrt{5^2 + (-6.5)^2} = \sqrt{25 + \frac{169}{4}}$

$$= \sqrt{\frac{269}{4}} = \frac{\sqrt{269}}{2}$$

This is the exact answer. Convert into a decimal to obtain the degree of accuracy asked for.

Length $= 8.20$ (2 d.p.)

The **midpoint** of a line, where the two end coordinates are known, can be found by adding the $x$-coordinates and dividing by 2 to find the $x$-coordinate of the midpoint, and adding the $y$-coordinates and dividing by 2 to get the $y$-coordinate. That is, the midpoint of the coordinates $(x_1, y_1)$ and $(x_2, y_2)$ is $\left(\frac{x_1 + x_2}{2}, \frac{y_1 + y_2}{2}\right)$.

## Example

Find the midpoint of the line CD, where C is $(-2, 5)$ and D is $(3, -15)$.

$x$-coordinate of midpoint $= \frac{-2 + 3}{2} = \frac{1}{2}$

$y$-coordinate of midpoint $= \frac{5 + -15}{2} = -5$

The midpoint is $\left(\frac{1}{2}, -5\right)$.

## Circles

● A tangent to a circle meets the radius at right angles. This means that the equations of the two lines are perpendicular so the gradients will be negative reciprocals of each other.
● Lines from either end of a diameter line will meet on the circumference of the circle at right angles.
● Radii are all equal length.
● A radius that meets a chord at a right angle will bisect the chord.

Equations of circles can be written in the form $(x - a)^2 + (y - b)^2 = r^2$. This form can be used to find the centre of the circle, $(a, b)$, and the radius, $r$.

If an equation is given for a circle but it is not in this form, using the method of **completing the square** will help to do so.

## Example

Find the centre and the radius of $y^2 + x^2 + 4y - 8x + 3 = 0$.

By grouping the $y$-terms close together, it is easier to complete the square. Ensure you show your method in clear, short steps.

$$y^2 + 4y + x^2 - 8x + 3 = 0$$

$$(y + 2)^2 - 4 + (x - 4)^2 - 16 + 3 = 0$$

Remember that to complete the square, both the $x^2$ and $y^2$ terms must have a coefficient of 1.

Collect the numbers together and move to the other side of the equation.

$$(y + 2)^2 + (x - 4)^2 = 17$$

Centre of the circle is $(4, -2)$; radius is $\sqrt{17}$.

Note: You take the positive square root, as you are talking about a distance.

## Parametric Equations

Sometimes a third parameter is used to define a relationship between two variables $x$ and $y$ (often $t$ or $\theta$). Parametric equations can describe quite complicated curves.

## Example

$y = 3\cos^2\theta - 2\cos\theta$ and $x = 2\sin^2\theta - 1$

Neither variable is defined in terms of the other but the two equations allow coordinates to be found using the third variable (parameter). Tables are a good way of organising this information:

| $\theta$ | $-\dfrac{5\pi}{6}$ | $-\dfrac{\pi}{6}$ | $\dfrac{\pi}{2}$ | $\dfrac{\pi}{3}$ | $\dfrac{2\pi}{3}$ | $\pi$ | $2\pi$ |
|---|---|---|---|---|---|---|---|
| $x$ | $-\dfrac{1}{2}$ | $-\dfrac{1}{2}$ | $1$ | $\dfrac{1}{2}$ | $\dfrac{1}{2}$ | $-1$ | $-1$ |
| $y$ | $\dfrac{9+4\sqrt{3}}{4}$ | $\dfrac{9-4\sqrt{3}}{4}$ | $0$ | $-\dfrac{1}{4}$ | $\dfrac{7}{4}$ | $5$ | $1$ |

All these coordinates would lie on the graph described by the parametric equations. Sometimes values may be given in a table from which the other values need to be found. Care must be taken about which points join to which. The graph would look like this:

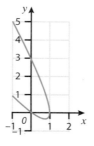

You can also convert a parametric equation into a Cartesian equation. To do so, express the parameter (or a term involving the parameter) in terms of one of the main variables and substitute it into the other equation. In the next example, trigonometric identities are important.

## Example

If converting $y = 3\cos^2\theta - 2\cos\theta$ and $x = 2\sin^2\theta - 1$ into a Cartesian equation, you can start with either equation. However, if you want $y$ as the subject, it is often simpler to rearrange the $x$ equation:

$x = 2\sin^2\theta - 1$

$x + 1 = 2\sin^2\theta$

$\sin^2\theta = \dfrac{x+1}{2}$

Using the identity $\sin^2\theta + \cos^2\theta = 1$:

$1 - \cos^2\theta = \dfrac{x+1}{2}$

$\cos^2\theta = 1 - \dfrac{x+1}{2} = \dfrac{1-x}{2}$

$\cos\theta = \pm\sqrt{\dfrac{1-x}{2}}$

$y = 3\cos^2\theta - 2\cos\theta$

Substituting in the values for $\cos\theta$ and $\cos^2\theta$ as found above:

$y = 3 \times \dfrac{1-x}{2} \pm 2 \times \sqrt{\dfrac{1-x}{2}}$

The final form needs to be as required by the question. If it isn't specified, any form is acceptable but look to simplify where possible.

$y = \dfrac{3-3x}{2} \pm \sqrt{2-2x}$

### Links to Other Concepts
- Straight lines describe all directly proportional relationships between simple variables
- Solving simultaneous equations to find the intersection points of two lines
- Transformations and functions
- Straight-line motion in mechanics
- Calculus
- Trigonometric identities

## SUMMARY

- If both variables are to the power 1 in an equation, the graph is a straight line.
- The gradient of a graph, $m = \frac{y_2 - y_1}{x_2 - x_1}$:
  - Parallel lines have equal gradients.
  - Perpendicular lines have negative reciprocal gradients, $m_2 = -\frac{1}{m_1}$ or $m_1 \times m_2 = -1$.
- The equation of a straight line is $y = mx + c$, $y - y_1 = m(x - x_1)$ or $ax + by + c = 0$.
- To find the midpoint of a line segment, add the end points and divide by 2.
- To find the length of a line, use Pythagoras' Theorem.

- The general equation of a circle is $(x - a)^2 + (y - b)^2 = r^2$, where the circle has centre $(a, b)$ and radius $r$. To get an equation into this form, use completing the square.
- To find the intersection points of two graphs, treat them as simultaneous equations.
- Parametric equations describe a function in terms of a third parameter. By substituting in values of the third parameter, you can find coordinate pairs for $x$ and $y$.
- By rearranging the parametric equations, you can often find the equation in Cartesian form.
- Trigonometric identities are often the key in converting between parametric and Cartesian equations.

## QUICK TEST

1. Match the parametric equations A–D to their graphs E–H:

   **A** $x = 2t, y = t^2 - 3$    **B** $x = 2\cos t, y = 2\sin t$

   **C** $x = 3t, y = t^2 - 2$    **D** $x = 2 - 5t, y = \frac{2}{t} + 2$

   **E**           **F**

   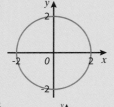 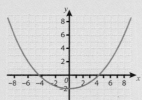

   **G**           **H**

   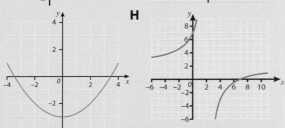

2. Find the midpoint and the length of the line segment that joins the points $(-3, 7)$ and $(2, -5)$.
3. Without using calculus, find the gradient of the tangent to the circle $(x + 10)^2 + (y - 12)^2 = 121$ at the point $(-1, 12 + 2\sqrt{10})$. Give the answer as an exact fraction with a rational denominator.

## PRACTICE QUESTIONS

1. The graph shows the curve $x = \tan\theta - 3$, $y = \tan 2\theta$

   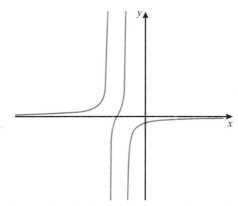

   a) Show that the Cartesian equation of the curve can be written as $y = -\frac{px + 2 + q}{(x + p)(x + q)}$   **[5 marks]**

   b) Hence or otherwise, find the equations of the asymptotes and the coordinates of axis-intercepts for the graph.   **[3 marks]**

2. A tangent meets a circle at point P $(1, 5)$ and has an equation $4x + 3y - 19 = 0$. The length of the line joining P to the centre of the circle is 5. Find the two possible equations for the circle.   **[8 marks]**

# Trigonometry 1

## The Basics

Trigonometry questions will be in both degrees and radians, so calculators will need to be adjusted accordingly. The basic trigonometric ratios are:

$$\sin\theta = \frac{\text{opposite}}{\text{hypotenuse}} \quad \cos\theta = \frac{\text{adjacent}}{\text{hypotenuse}} \quad \tan\theta = \frac{\text{opposite}}{\text{adjacent}}$$

In addition to the basic trigonometric functions, the **reciprocal trigonometric functions** are used:

$$\textbf{cosec}\ \theta = \frac{1}{\sin\theta} = \frac{\text{hypotenuse}}{\text{opposite}}$$

$$\textbf{sec}\ \theta = \frac{1}{\cos\theta} = \frac{\text{hypotenuse}}{\text{adjacent}}$$

$$\textbf{cot}\ \theta = \frac{1}{\tan\theta} = \frac{\text{adjacent}}{\text{opposite}}$$

Also included at A-level are the **inverse functions**. They effectively find the angle given a ratio of sides:

$$\textbf{arcsin}\,x = \sin^{-1}x \quad \textbf{arccos}\,x = \cos^{-1}x \quad \textbf{arctan}\,x = \tan^{-1}x$$

## Non-Right-Angled Triangles

The following trigonometric equations can be used with non-right-angled triangles.

### Area of Triangle $= \frac{1}{2}ab\sin C$

This equation is used for any question that involves area. It is set up to find the **area** given two sides and the angle between them. It can be rearranged to find a side or an angle if the area is given.

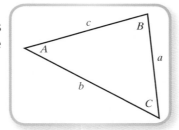

---

### The Sine Rule: $\frac{a}{\sin A} = \frac{b}{\sin B} = \frac{c}{\sin C}$

The sine rule can also be written $\frac{\sin A}{a} = \frac{\sin B}{b} = \frac{\sin C}{c}$. This equation is used when given an angle–side pair and then either an angle or a side, and you are asked to find the **corresponding** angle or side (i.e. when two angles and two sides are involved, one of which is the value to be found).

Use with **internal angles** of a triangle (which add up to 180°) to create the required angle–side pair if needed.

The ambiguous case: It is possible if given angle $A$, side $a$ and side $c$ (i.e. SSA, two sides and an angle that isn't between them) that there are two possible positions for the line BC, as shown in the diagram.

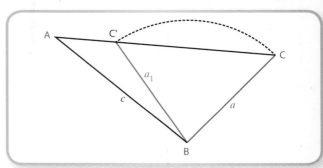

The question may hint at which is the correct one but it is important to acknowledge both then reject one with a reason, or give both as a final answer if they are relevant.

By recognising an **SSA** case and finding two possible answers, you can check if both are valid by making sure that the sum of the obtuse angle found and the original angle given is less than 180°. Calculators will only give one result, $\theta$. To find the second result, use $180 - \theta$. You should expect two answers to be valid if given $S_1 S_2 A$, where $S_1$ is opposite angle $A$, such that $S_1 < S_2$.

---

### Example

Given that the area of triangle ABC is 12 and sides $AB = 4$ and $AC = 7$, find angle $A$ to 1 decimal place.

$$12 = \frac{1}{2} \times 4 \times 7 \times \sin A$$

$$\sin A = \frac{12}{\frac{1}{2} \times 4 \times 7} = \frac{6}{7}$$

$$A = \sin^{-1}\left(\frac{6}{7}\right) = 58.9972\ldots = 59.0° \text{ (1 d.p.)}$$

## Example

Triangle PQR has sides PR = 11 cm, RQ = 6 cm and angle $P = 27°$.

**a)** Find the size of angle $Q$, giving your answer to 1 decimal place.

Sketch the triangle:

Use the sine rule:

$$\frac{\sin Q}{11} = \frac{\sin 27}{6}$$

$$\sin Q = \frac{11 \sin 27}{6}$$

$Q = 56.3373\ldots$ (store in calculator as A to retain accuracy)

As SSA case, is there a second result?

$180 - 56.337\ldots = 123.66\ldots$ (store in calculator as B)

$27 + 123.66\ldots < 180$ ∴ both results are valid.

$Q = 56.3°$ or $123.7°$ (1 d.p.)

**b)** The perimeter of the shape is less than 28 cm. Use the sine rule to find the size(s) of side PQ.

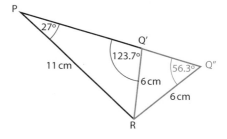

The two options for the position of Q are shown.

If perimeter < 28 cm, then
PQ < 28 − (11 + 6) = 11 cm

Using triangle PRQ':

$$\frac{11}{\sin 123.7} = \frac{PQ'}{\sin(180 - 27 - 123.7)}$$

$$\Rightarrow PQ' = \frac{11}{\sin 123.7} \times \sin 29.3 = 6.47$$

Using triangle PRQ":

Q'R = Q"R = 6 cm, so triangle RQ'Q" is isosceles and

Q'Q" = 2(6 cos 56.3) = 6.66 cm and
PQ" = 6.47 + 6.66 = 13.1 cm > 11 cm

Therefore PQ = 6.47 cm

---

## The Cosine Rule: $a^2 = b^2 + c^2 - 2bc \cos A$

The cosine rule is used to find an unknown side when given a pair of sides and the angle between them. It can also be rearranged to find $b$, $c$ or $A$. Finding $b$ or $c$ will involve solving a quadratic. The cosine rule is used if only one angle is involved, whether it is the unknown or a given, i.e. when three sides and an angle are involved in the question, one of which is the unknown.

### Example

A triangle ABC has an area of 15 cm². Side AC = 5 cm and angle $C = 60°$. Find the length of line AB, to 3 significant figures.

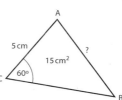

Start by finding a side using the area rule:

$$15 = \frac{1}{2} \times 5 \times BC \times \sin 60$$

$$BC = \frac{15}{\frac{1}{2} \times 5 \times \sin 60} = 4\sqrt{3} \text{ cm}$$

To find AB, now apply the cosine rule (as there is only one angle): $a^2 = b^2 + c^2 - 2bc \cos A$

$$AB^2 = (4\sqrt{3})^2 + 5^2 - 2 \times 5 \times 4\sqrt{3} \times \cos 60$$

$$= 73 - 20\sqrt{3}$$

$$AB = \sqrt{73 - 20\sqrt{3}} = 6.1934\ldots = 6.19 \text{ cm (3 s.f.)}$$

---

## Radian Measure

Radians are a way of measuring angles. 1 radian is the angle subtended by an arc that is equal to the radius:

| $360° = 2\pi$ radians | $180° = \pi$ radians | $90° = \frac{\pi}{2}$ radians |
|---|---|---|

Arc length: $s = r\theta$ 　　　Area of sector: $A = \frac{1}{2}r^2\theta$

### Example

The diagram shows a sector of a circle with an angle of 240°.

**a)** Convert 240° to radians.

$$\frac{240}{360} \times 2\pi = \frac{4\pi}{3} \text{ radians}$$

**b)** Find the shape's perimeter.

$$s = r\theta = 6 \times \frac{4\pi}{3} = 8\pi \text{ cm}$$

Perimeter = 2r + arc length

$$= 2 \times 6 + 8\pi = 12 + 8\pi \text{ cm}$$

## Graphs of Sine, Cosine and Tangent

While the roots and proofs of trigonometric ratios lie in triangles, remember that sine, cosine and tangent are many-to-one functions. This means you can look at angles beyond the 180° in a triangle. For any given ratio, there are many possible angles that would give the required result. However, each $x$-value only gives one $y$-value. The key features of the sine, cosine and tangent graphs are shown:

$y = \sin x$

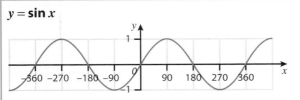

- Maximum value 1, minimum value −1 and $y$-intercept $= 0$
- Vertical lines of symmetry through every minimum and maximum.
  $\ldots, -270°, -90°, 90°, 270°, \ldots$ or $\ldots, -\dfrac{3\pi}{2}, -\dfrac{\pi}{2}, \dfrac{\pi}{2}, \dfrac{3\pi}{2}, \ldots$
- Rotational symmetry about the origin $\rightarrow \sin(\alpha) = -\sin(-\alpha)$
- Periodicity 360° or $2\pi$ radians $\rightarrow \sin(\beta) = \sin(\beta + 360n)$ or $\sin(\beta + 2\pi n)$, where $n$ is an integer value

$y = \cos x$

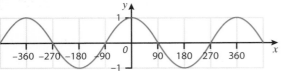

- Maximum value 1, minimum value −1 and $y$-intercept $= 1$
- Vertical lines of symmetry through every minimum and maximum.
  $\ldots, -360°, -180°, 0°, 180°, 360°, \ldots$ or $\ldots, -2\pi, -\pi, 0, \pi, 2\pi, \ldots$
- Reflectional symmetry about the $y$-axis $\rightarrow \cos(\alpha) = \cos(-\alpha)$
- Periodicity 360° or $2\pi$ radians $\rightarrow \cos(\beta) = \cos(\beta + 360n)$ or $\cos(\beta + 2\pi n)$, where $n$ is an integer value

$y = \tan x$

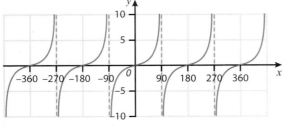

- Vertical asymptotes at $\ldots, -270°, -90°, 90°, 270°, \ldots$ or $\ldots, -\dfrac{3\pi}{2}, -\dfrac{\pi}{2}, \dfrac{\pi}{2}, \dfrac{3\pi}{2}, \ldots$
- Periodicity 180° or $\pi$ radians $\rightarrow \tan(\beta) = \tan(\beta + 180n)$ or $\tan(\beta + \pi n)$, where $n$ is an integer value
- $y$-intercept $= 0$
- Rotational symmetry of order 2 about the $x$-intercepts $\rightarrow \tan(\alpha) = -\tan(-\alpha)$
- No lines of symmetry

The graphs of the **reciprocal functions** are:

$y = \mathbf{cosec}\, x$

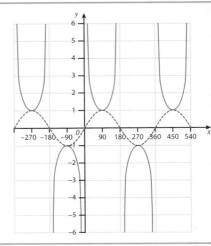

$y = \mathbf{sec}\, x$

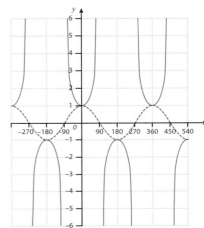

$y = \mathbf{cot}\, x$

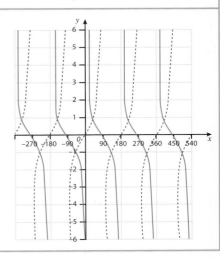

The graphs of the **inverse functions** are a reflection of $y = \sin x$, $y = \cos x$ and $y = \tan x$ respectively, reflected in the line $y = x$.

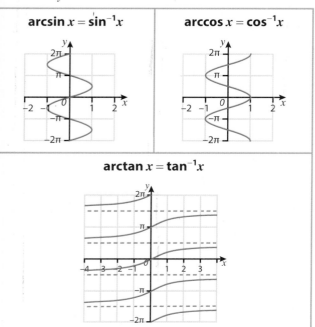

| arcsin $x = \sin^{-1}x$ | arccos $x = \cos^{-1}x$ |
| --- | --- |

arctan $x = \tan^{-1}x$

When solving a trigonometric equation, using these properties of the graphs allows you to find all the possible solutions within a given range. Sketching the relevant graph can help to identify the number of solutions and how to find them.

**Example**

Find all the solutions to $\sin^2(x) = 0.25$ in the range $0° \leqslant x \leqslant 360°$.

$\sin(x) = \pm\sqrt{0.25} = \pm 0.5$

$\sin^{-1}(0.5) = 30°$

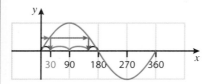

Calculator result 30°

Based on symmetry about $x = 90$, there is an additional result at $\boxed{180 - 30} \rightarrow 150 \leftarrow \boxed{90 + 60}$

$\sin^{-1}(-0.5) = -30°$, which is outside the required range but can be used to find the solutions within the range.

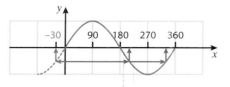

Calculator result $-30°$

Additional results of: $180 + 30 = 210°$

$\qquad\qquad\qquad\qquad -30 + 360 = 330°$

$x = 30°, 150°, 210°, 330°$

Note: Substituting results back into a calculator is a way of double checking them, e.g. $\sin(210) = -0.5$.

## Small Angle Approximations

With small angles, measured in radians, you can use these approximations:

$$\sin x \approx x \qquad \tan x \approx x \qquad \cos x \approx 1 - \frac{1}{2}x^2$$

**Example**

a) Using a small angle approximation, show that $7\sin x - 4\cos x \approx (2x-1)(x+4)$, given that $x \leqslant \frac{\pi}{50}$.

$$7\sin x - 4\cos x \approx 7x - 4\left(1 - \frac{1}{2}x^2\right)$$

$$\approx 7x - 4 + 2x^2$$

$$\approx 2x^2 + 7x - 4$$

$$\approx (2x-1)(x+4)$$

b) What is the maximum percentage error if using this approximation? Give your answer to 3 significant figures.

Maximum value of $x = \frac{\pi}{50}$

$$\frac{\left(7\sin\left(\frac{\pi}{50}\right) - 4\cos\left(\frac{\pi}{50}\right)\right) - \left(2 \times \frac{\pi}{50} - 1\right)\left(\frac{\pi}{50} + 4\right)}{7\sin\left(\frac{\pi}{50}\right) - 4\cos\left(\frac{\pi}{50}\right)} \times 100$$

$$= 0.00822\% \text{ (3 s.f.)}$$

Note: If you get an answer of ~11%, check your calculator is set to radians. Given that $\frac{\pi}{50}$ is a sufficiently small angle, the percentage error should be relatively small. If your result doesn't feel right, check your working and your calculator.

## Known Angles

You should know certain values of sine, cosine and tangent. A calculator can find these values but being able to recognise them yourself is very useful. Some calculators will return the exact value (in surd form) but others may return a decimal value, which would create a rounding error.

These two triangles show where the values of the trigonometric functions come from:

From knowing $\sin 30 = \frac{1}{2}$:

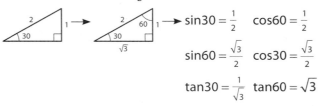

$$\sin 30 = \frac{1}{2} \quad \cos 60 = \frac{1}{2}$$

$$\sin 60 = \frac{\sqrt{3}}{2} \quad \cos 30 = \frac{\sqrt{3}}{2}$$

$$\tan 30 = \frac{1}{\sqrt{3}} \quad \tan 60 = \sqrt{3}$$

Right-angled triangle with 45° is isosceles:

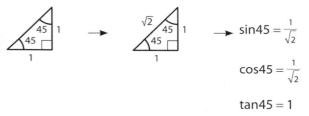

$$\sin 45 = \frac{1}{\sqrt{2}}$$

$$\cos 45 = \frac{1}{\sqrt{2}}$$

$$\tan 45 = 1$$

### Links to Other Concepts

- Bearings, triangles, coordinate geometry
- 'Hidden' quadratics giving a positive and negative result for a trigonometric ratio
- Transformations of the sine, cosine and tangent graphs
- Circles and lines, finding angles within a circle
- Proof
- Pythagoras' Theorem where right-angled triangles are involved
- Vectors and mechanics
- Trigonometric identities
- Calculus

## SUMMARY

- Area of a triangle $= \frac{1}{2}ab \sin C$
- The sine rule: $\frac{a}{\sin A} = \frac{b}{\sin B} = \frac{c}{\sin C}$
- Watch out for SSA cases when applying the sine rule. Check for two possible results.
- The cosine rule: $a^2 = b^2 + c^2 - 2bc \cos A$
- $\theta$, $\alpha$ and $\beta$ are often used to denote angles.
- If dividing through by a trigonometric function which is a common factor in all terms of the equation, that function equalling zero provides a valid result that can be easily forgotten.
- Radians are used to measure angles. Any questions that involve calculus should be in radians.
- $360° = 2\pi$ radians
- Arc length: $s = r\theta$
- Area of sector: $A = \frac{1}{2}r^2\theta$
- Small angle approximations, when angles are measured in radians:

  $\sin x \approx x$

  $\tan x \approx x$

  $\cos x \approx 1 - \frac{1}{2}x^2$

- You are expected to know the values for sine, cosine and tangent for 30°, 45° and 60° (or the equivalent radian measures).
- Ensure your calculator is in the correct mode (degrees or radians) for questions involving trigonometric ratios.

1. Convert these measures from degrees to radians, or radians to degrees.

   **a)** $24°$

   **b)** $\frac{4\pi}{5}$ radians

   **c)** $1260°$

   **d)** $1.06$ radians

2. Find the value of $r$.

3. Find the value of angle $\theta$. Give your answer in both degrees and radians, accurate to 3 significant figures.

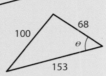

4. The diagram shows a sector of a circle.

   **a)** Find the area of the sector.

   **b)** Find the perimeter of the sector.

5. Without using a calculator, evaluate $\cos\left(\frac{4\pi}{3}\right)\left(\tan\left(\frac{5\pi}{4}\right) - m\cos\left(-\frac{\pi}{6}\right)\right)$.

   Express the answer as a single fraction.

6. Approximate the following expressions using the small angle approximations.

   **a)** $\sin 7.5°$

   **b)** $\cos 0.02 - \tan 0.01$ (angles measured in radians)

   **c)** What is the percentage error in part **a)**?

7. Without using a calculator, evaluate the following:

   **a)** $\cos(\arctan 1)$

   **b)** $\sin\left(\arccos\left(\frac{\sqrt{3}}{2}\right)\right)$

   **c)** $\operatorname{cosec}\left(\frac{\pi}{2}\right)$

   **d)** $\cot(30°)$

   **e)** $\sec(30°) \times \tan\left(\arcsin\left(\frac{1}{2}\right)\right)$

## PRACTICE QUESTIONS

1. $(3\cos x - 1)(5\sin x - 2)\tan x = 0$. Find all the possible values of $x$ in the range $-\pi \leqslant x \leqslant 2\pi$. **[4 marks]**

2. Find the area of the composite shape ABCDE formed of three triangles. **[7 marks]**

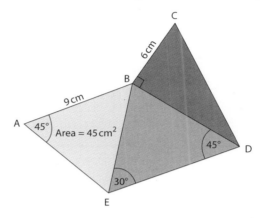

   Angle BAE $= 45°$

   Angle BED $= 30°$

   Angle BDE $= 45°$

   AB $= 9\,$cm and BC $= 6\,$cm

# Trigonometry 2

## Trigonometric Identities

You may need to use trigonometric identities to simplify an equation to the point where it can be solved using the methods met previously.

$$\frac{\sin\theta}{\cos\theta} = \frac{\frac{\text{opp}}{\text{hyp}}}{\frac{\text{adj}}{\text{hyp}}} = \frac{\text{opp}}{\text{adj}} \rightarrow \mathbf{\tan\ \theta \equiv \frac{\sin\ \theta}{\cos\ \theta}}$$

Also note that $\mathbf{\tan^2\theta \equiv \frac{\sin^2\theta}{\cos^2\theta}}$

### Example

$4\cos\theta\tan\theta = 1$. Find all the solutions to $\theta$ in the range $0° < \theta \leqslant 360°$ accurate to 1 decimal place.

Substitute in $\tan\theta \equiv \frac{\sin\theta}{\cos\theta}$

$4\cos\theta\frac{\sin\theta}{\cos\theta} = 1$

$4\sin\theta = 1$

$\sin\theta = \frac{1}{4}$

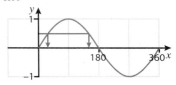

$\theta = 14.4775\ldots = 14.5°$ (1 d.p.)

$\theta = 180 - 14.4775\ldots = 165.522\ldots° = 165.5°$ (1 d.p.)

### $\sin^2\theta + \cos^2\theta \equiv 1$

$\sin^2\theta + \cos^2\theta \equiv 1$ can be rearranged to give the following identities:

$\mathbf{\sin^2\theta \equiv 1 - \cos^2\theta}$ and $\mathbf{\cos^2\theta \equiv 1 - \sin^2\theta}$

By dividing by $\sin^2\theta$: $\mathbf{1 + \cot^2\theta \equiv \csc^2\theta}$

By dividing by $\cos^2\theta$: $\mathbf{\tan^2\theta + 1 \equiv \sec^2\theta}$

Rearranging might be used to simplify an equation or to remove a function so that it can be solved. It could also be used to find the corresponding sine and cosine value given one or the other, in its exact form.

### Example

Given that $\sin\theta = \frac{2}{5}$, and $\theta$ is an acute angle:

a) Find the exact value of $\cos\theta$.

---

By finding $\theta$, then substituting into cos, the answer from a calculator would be 0.91651… Another method is needed to find this as an exact value.

Using the identity $\cos^2\theta \equiv 1 - \sin^2\theta$

$\cos^2\theta = 1 - \left(\frac{2}{5}\right)^2$

$\cos^2\theta = \frac{21}{25}$

$\cos\theta = \sqrt{\frac{21}{25}} = \frac{\sqrt{21}}{5}$

b) Hence or otherwise, find the value of $\tan\theta$.

$\tan\theta = \frac{\sin\theta}{\cos\theta} = \frac{\frac{2}{5}}{\frac{\sqrt{21}}{5}}$

$\tan\theta = \frac{2}{\sqrt{21}} = \frac{2\sqrt{21}}{21}$

As the question states that $\theta$ is acute, both $\tan\theta$ and $\cos\theta$ will be positive.

c) If $\theta$ was obtuse instead of acute, what would the value of $\cos\theta$ be?

Looking at the graphs of sine and cosine together, you can see that the value of $\cos\theta$ will now be negative but of the same magnitude:

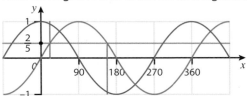

$\cos\theta = -\frac{\sqrt{21}}{5}$

Using the CAST diagram would give the same result and is some people's preferred method:

Acute – all positive

$\cos\theta = \frac{\sqrt{21}}{5}$

$\tan\theta = \frac{2}{\sqrt{21}} = \frac{2\sqrt{21}}{21}$

Obtuse – only sine positive

$\cos\theta = -\frac{\sqrt{21}}{5}$

$\tan\theta = -\frac{2}{\sqrt{21}} = -\frac{2\sqrt{21}}{21}$

Look out for when the equation forms a quadratic. Remember both positive and negative values when taking a square root. Sometimes both answers will be valid. At other times, an answer can be discounted as it is too large or too small (e.g. $\cos\theta = 2$, since cosine has a maximum value of 1 and a minimum of $-1$).

It can be hard to know where to start with trigonometric identity questions. Key things to look out for:

● Mixed trigonometric functions – is it possible to get the equation in terms of just one?
● Trigonometric functions in denominators – if it makes it look more complicated, consider multiplying through by the denominator to see what happens.
● What is the bit of the equation that seems most complicated and is there a way to simplify it (multiplying out brackets, multiplying up by denominators, etc.)?

---

**Example**

Given that $3\tan\theta = 2\cos\theta$:

**a)** Show that $2\sin^2\theta + 3\sin\theta - 2 = 0$.

$$3\tan\theta = 2\cos\theta$$
$$3\frac{\sin\theta}{\cos\theta} = 2\cos\theta$$
$$3\sin\theta = 2\cos^2\theta$$

Multiplying up by cosine creates cosine-squared, for which we have another identity:

$$3\sin\theta = 2(1 - \sin^2\theta)$$
$$3\sin\theta = 2 - 2\sin^2\theta$$
$$2\sin^2\theta + 3\sin\theta - 2 = 0$$

**b)** Hence or otherwise, find the solutions for $\theta$ in the range $-180° < \theta \leqslant 180°$.

$$2\sin^2\theta + 3\sin\theta - 2 = 0$$
$$(2\sin\theta - 1)(\sin\theta + 2) = 0$$

Factorisation is great for simple quadratics but completing the square or the quadratic formula are just as good.

$\sin\theta + 2 = 0 \Rightarrow \sin\theta = -2$, which is impossible

Acknowledging a result is important, even if you disregard it.

$$2\sin\theta - 1 = 0 \Rightarrow \sin\theta = \frac{1}{2}$$
$$\sin^{-1}\left(\frac{1}{2}\right) = 30°$$
$$\theta = 30°, 150°$$

---

## Compound Angle Formulae

Compound angle formulae allow you to manipulate equations involving trigonometric functions with different angles. All the previous identities rely on the angles within the functions being the same.

● $\sin(A \pm B) \equiv \sin A \cos B \pm \cos A \sin B$

● $\cos(A \pm B) \equiv \cos A \cos B \mp \sin A \sin B$

● $\tan(A \pm B) \equiv \dfrac{\tan A \pm \tan B}{1 \mp \tan A \tan B}$

The **double angle formulae** can be derived from the compound angle formulae:

● $\sin 2A \equiv 2\sin A \cos A$

● $\cos 2A \equiv \cos^2 A - \sin^2 A \equiv 1 - 2\sin^2 A \equiv 2\cos^2 A - 1$

● $\tan 2A \equiv \dfrac{2\tan A}{1 - \tan^2 A}$

Note: Sometimes questions will specify whether they are in degrees or radians, but at other times you are expected to use clues in the question. A range given in terms of $\pi$ suggests radians.

---

**Example**

Given that $\cos\left(\theta + \dfrac{\pi}{6}\right) = 3\sin\theta$, and $\theta$ is a reflex angle, find the value of $\theta$.

$$\cos\frac{\pi}{6}\cos\theta - \sin\frac{\pi}{6}\sin\theta = 3\sin\theta$$

$$\frac{\sqrt{3}}{2}\cos\theta - \frac{1}{2}\sin\theta = 3\sin\theta$$

$$\frac{\sqrt{3}}{2}\cos\theta = \frac{7}{2}\sin\theta$$

$$\frac{2 \times \sqrt{3}}{7 \times 2}\cos\theta = \sin\theta$$

$$\tan\theta = \frac{2\sqrt{3}}{14}$$

$$\theta = \arctan\frac{2\sqrt{3}}{14} = 0.2425638\ldots$$

As the angle is reflex, add $\pi$:
$\theta = 0.24256\ldots + \pi = 3.38$ radians (3 s.f.)

---

## Equivalent Form $\cos(\theta + \alpha)$

A function $f(x) = a\cos\theta \pm b\sin\theta$ (where $a$ and $b$ are positive values) can be presented in these forms:

- $a\sin\theta \pm b\cos\theta \equiv r\sin(\theta \pm \alpha)$
- $a\cos\theta \pm b\sin\theta \equiv r\cos(\theta \mp \alpha)$

The compound angle formulae can be used to expand the expressions and, hence, coefficients can be compared. Take care to use + and − correctly.

---

**Example**

Express $3\cos\theta + 4\sin\theta$ in the form $r\cos(\theta - \alpha)$, where $r$ and $\alpha$ are values to be found.

Expanding $r\cos(\theta - \alpha) = r(\cos\theta\cos\alpha + \sin\theta\sin\alpha)$

$$= (r\cos\alpha)\cos\theta + (r\sin\alpha)\sin\theta$$

By comparing coefficients:

$3 = r\cos\alpha$

$4 = r\sin\alpha$

$\dfrac{r\sin\alpha}{r\cos\alpha} = \tan\alpha = \dfrac{4}{3}$

$\alpha = \tan^{-1}\left(\dfrac{4}{3}\right)$

$\alpha = 53.1°$ (3 s.f.)

$r = \dfrac{3}{\cos 53.1\ldots} = 5$

$3\cos\theta + 4\sin\theta = 5\cos(\theta - 53.1)$

---

Equivalent form can be used, along with knowledge of graph transformations, to find details such as the period, maxima, minima and axis-intercepts of the graphs.

---

**Example**

**a)** Use the form $r\sin(\theta - \alpha)$ to sketch the curve $y = 24\sin x - 7\cos x$, clearly stating the values of the maximum, minimum, $x$-intercepts and the $y$-intercept.

$r\sin(\theta - \alpha) = r(\sin\theta\cos\alpha - \cos\theta\sin\alpha)$

$24 = r\cos\alpha$ and $7 = r\sin\alpha$

$\tan\alpha = \dfrac{7}{24}$

$\alpha = \arctan\dfrac{7}{24} = 16.26\ldots$

---

$r = \dfrac{24}{\cos 16.26\ldots} = 25$

$y = 25\sin(x - 16.26\ldots°)$

To find the maximum, the graph is a vertical stretch of 25. The maximum of $\sin x$ is 1 so the maximum in this case is 25 and the minimum is −25.

There is a horizontal translation of $16.26\ldots°$ to the right so the $x$-intercepts are at $16.26\ldots$, $180 + 16.26\ldots$, etc. The $y$-intercept can be found using $y = 25\sin(-16.26\ldots)$

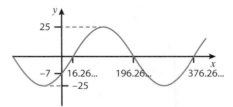

**b)** Hence or otherwise, describe the combination of transformations that map $y = \cos x$ onto $y = 24\sin x - 7\cos x + 7$ and state the value of the $y$-intercept.

The curve $y = \cos x$ experiences a stretch with scale factor 25 vertically.

The curve is translated $\binom{106.26}{7}$

As the graph repeats, any horizontal component for the translation in the form $360n + 90 + 16.26\ldots$, where $n \in \mathbb{Z}$, is acceptable. For example, $-253.74$ (2 d.p.) as it is $360 \times -1 + 106.26\ldots$

The $y$-intercept is $y = 0$

---

**Links to Other Concepts**

- Surds
- Bearings, triangles, coordinate geometry
- 'Hidden' quadratics giving a positive and a negative result for a trigonometric ratio
- Transforming the sine, cosine and tangent graphs
- Proof, justification and algebraic manipulation
- Mechanics, resolving forces
- Calculus

- Trigonometric identities are used to simplify expressions and convert between the different functions:

$$\tan\theta \equiv \frac{\sin\theta}{\cos\theta} \qquad \cot\theta \equiv \frac{\cos\theta}{\sin\theta}$$

$$\sin^2\theta + \cos^2\theta \equiv 1$$

$$\tan^2\theta + 1 \equiv \sec^2\theta$$

$$1 + \cot^2\theta \equiv \operatorname{cosec}^2\theta$$

- Compound angle formulae:

$$\sin(A \pm B) \equiv \sin A \cos B \pm \cos A \sin B$$

$$\cos(A \pm B) \equiv \cos A \cos B \mp \sin A \sin B$$

$$\tan(A \pm B) \equiv \frac{\tan A \pm \tan B}{1 \mp \tan A \tan B}$$

**Watch out for $\pm$ and $\mp$.**

- Double angle formulae can be derived from the compound angle formulae:

$$\sin 2A \equiv 2 \sin A \cos A$$

$$\cos 2A \equiv \cos^2 A - \sin^2 A$$

$$\equiv 1 - 2\sin^2 A \equiv 2\cos^2 A - 1$$

$$\tan 2A \equiv \frac{2 \tan A}{1 - \tan^2 A}$$

- Sketches of the functions will help to find all the results in a given range.
- Exact results will generally contain surds. If asked for an exact result, a decimal answer from a calculator is not acceptable (but it could be used as a check).
- When applying identities, the angle must be exactly the same.
- When solving an equation where the angle has a multiplier, the range needs to be adjusted or the transformed graph used in order to get all the results.
- The 'R formula' can be used to rewrite equations in an equivalent form:

$$a\sin\theta \pm b\cos\theta \equiv r\sin(\theta \pm \alpha)$$

$$a\cos\theta \pm b\sin\theta \equiv r\cos(\theta \mp \alpha)$$

1. Use the compound angle formulae to show that $\tan 2p \equiv \dfrac{2\tan p}{1 - \tan^2 p}$

2. Find the values for which $\cos x = -\sin 2x$ in the range $0 \leqslant x \leqslant 2\pi$.

3. Match A–F to their equivalent expressions, G–L:

| | | | | |
|---|---|---|---|---|
| **A** | $\sin^2 2\theta$ | **G** | $\cos^2\theta$ |
| **B** | $\sin(\theta + 30)$ | **H** | $\sin 6\theta \cos\theta - \cos 6\theta \sin\theta$ |
| **C** | $\sec^2\theta$ | **I** | $\dfrac{1 + \sin^2\theta}{1 - \sin^4\theta}$ |
| **D** | $\dfrac{1}{\tan^2\theta + 1}$ | **J** | $4\cos^2\theta - 4\cos^4\theta$ |
| **E** | $\sin 5\theta$ | **K** | $\dfrac{\sin\theta + \cos\theta}{\cos\theta - \sin\theta}$ |
| **F** | $\tan\left(\theta + \dfrac{\pi}{4}\right)$ | **L** | $\dfrac{1}{2}\cos\theta + \dfrac{\sqrt{3}}{2}\sin\theta$ |

## PRACTICE QUESTIONS

1. **a)** Show that

$$\tan^2 x + \frac{3}{1 - 2\sin^2\frac{x}{2}} - 9 \equiv \sec^2 x + 3\sec x - 10.$$

**[5 marks]**

**b)** Hence or otherwise, find the solutions to $\tan^2 x + \dfrac{3}{1 - 2\sin^2\frac{x}{2}} - 9 = 0$, where $0 < x < 2\pi$.

**[4 marks]**

2. **a)** $6\sin x - p\cos x$ can be expressed in the form $10\sin(\theta - \alpha)$. Find the values of $p$ and $\alpha$.

**[4 marks]**

**b)** Hence state for what values of $q$ the graph of $y = \dfrac{1}{6\sin x - p\cos x + q}$ will have no asymptotes.

**[5 marks]**

**c)** Hence find the exact values of the greatest and least values of $y$ for the curve $y = \dfrac{1}{6\sin x - p\cos x + 12}$

**[4 marks]**

# Trigonometry 3

## Finding All the Results

In some cases, simplifying a trigonometric (or algebraic) equation may involve dividing through by a common factor which is a function of $x$.

> ### Example
> $\cot x = 2\cos x$
> Solve for $x$ in the range $-\pi < x \leqslant 2\pi$.
>
> $\cos x \operatorname{cosec} x = 2\cos x$
>
> Both sides have a factor of $\cos x$, so you can find solutions to the equation by dividing by $\cos x$:
>
> $\operatorname{cosec} x = 2$
>
> $\sin x = \dfrac{1}{2}$
>
> $x = \dfrac{\pi}{6}, \dfrac{5\pi}{6}$   Note: These are expected to be known values.
>
> However, by simply cancelling the common factor, a set of results is potentially lost; this is the case when $\cos x = 0$.
>
> $\cos x = 0$
>
> $x = -\dfrac{\pi}{2}, \dfrac{\pi}{2}, \dfrac{3\pi}{2}$
>
> So the full set of results to the equation $\cot x = 2\cos x$ is $-\dfrac{\pi}{2}, \dfrac{\pi}{6}, \dfrac{\pi}{2}, \dfrac{5\pi}{6}, \dfrac{3\pi}{2}$

Using a graphics calculator to sketch the graphs and find the solutions, or the number of solutions, can be a good way to check results. If you do so, show what equation was entered, what the graph looked like (sketch) and how it was used to find the results.

> ### Example
> For $\cot x = 2\cos x$ above, plot $y = \cot x - 2\cos x$ or $y = 2\cos x - \cot x$ (shown here) and look at the number of $x$-intercepts.
>
> There are five $x$-intercepts in the given range. This matches with the results above.

## Multiples of the Unknown Angle

Using your knowledge of transformations of graphs, you can solve problems where the angle within the function has a multiplier.

$y = f(ax)$ gives a graph that is stretched horizontally by a factor of $\dfrac{1}{a}$.

There are different ways of approaching this type of question; two key methods are detailed below.

### Method 1: Extending the Range

Solve for $a\theta$ and find a set of results. The range stated in the question has to be adjusted so all the results for $\theta$ are found.

### Method 2: Transforming the Graph

By sketching the transformed graph, use its symmetry and periodicity to solve for $\theta$.

> ### Example
> Find all the solutions of $\cos 2x = 0.6$ in the range $-180° < x \leqslant 180°$.
>
> #### Method 1
> The range needs to be doubled to find all the values that, when halved, would fall into the range $-180° < x \leqslant 180°$.
>
> Let $\alpha = 2x$
>
> Solve $\cos \alpha = 0.6$ in the range $-360° < \alpha \leqslant 360°$.
>
>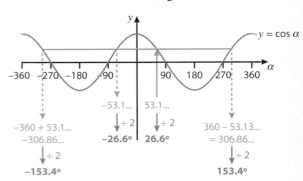

Remember to halve for the final answer.

## Method 2

The graph is squashed in by half. Having found one result for $x$, sketching the transformed graph allows the rest of the results to be found.

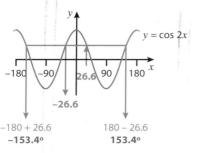

## Pulling it All Together

Questions will generally include more than one element. As with algebraic manipulations, look for a starting point and see what happens. Recognising when an equation can be simplified is important.

### Example

Given that $1 + \tan^2 3\theta = \frac{1}{\cos^2 3\theta}(7 - \tan 3\theta)$, find all the solutions for $\theta$ in the range $-180° < \theta \leqslant 180°$.

Let $\alpha = 3\theta$ and solve in the range $-540° < \alpha \leqslant 540°$.

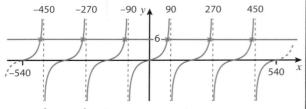

$1 + \tan^2\alpha = \frac{1}{\cos^2\alpha}(7 - \tan\alpha)$

$\cos^2\alpha + \cos^2\alpha \tan^2\alpha = 7 - \tan\alpha$

$\cos^2\alpha + \cos^2\alpha \frac{\sin^2\alpha}{\cos^2\alpha} = 7 - \tan\alpha$

$\cos^2\alpha + \sin^2\alpha = 7 - \tan\alpha$

$1 = 7 - \tan\alpha$

$\tan\alpha = 6$

$\alpha = \tan^{-1}(6) = 80.537677...°$

$\alpha = -459.4...°, -279.4...°, -99.4...°, 80.5...°,$
$260.5...°, 440.5...°$

$\alpha = 3\theta \rightarrow \theta = \frac{\alpha}{3}$

$\theta = -153.2°, -93.2°, -33.2°, 26.8°, 86.8°, 146.8°$
(answers to 1 d.p.)

### Links to Other Concepts
- Bearings, triangles, coordinate geometry
- 'Hidden' quadratics giving a positive and a negative result for a trigonometric ratio
- Transformations of graphs
- Mechanics, resolving forces  ● Calculus

## SUMMARY

- Sketches of the functions will help to find all the results in a given range.
- If finding an angle, remember to use a technique (transformed graph or adjusted range) to find all the relevant results.
- Quadratics need to have all results considered.
- When simplifying by dividing an equation through by a common factor which is a function of $x$, ($f(x)$), remember that some solutions would be found from $f(x) = 0$.

## QUICK TEST

1. How many solutions are there to the equation $3\sin^2(x) + 2\sin x = 0$ in the range $0 \leqslant x < 720°$?

2. Check the solutions for these equations for the range $0° \leqslant x \leqslant 360°$. Add any missing results.

   a) $4\sin^2(2x) = 1$
      Solutions $x = 75, 110, 160, 195, 255$

   b) $\tan^2(x - 20) = 7 - 4\sqrt{3}$
      Solutions $x = 15, 95, 185, 215$

   c) $\left(\cos\left(\frac{x}{2} + 15\right)\right)(4\sin^2 2x - 3) = 0$
      Solutions $x = 30, 60, 172.5, 210, 240$

## PRACTICE QUESTIONS

1. $\left(\tan^2\frac{\theta}{3} - 25\right)\left(5\sin\frac{\theta}{2}\right) = 0$
   Find all the solutions for $\theta$ in the range $-1000° < \theta < 1000°$, accurate to 3 significant figures.
   **[7 marks]**

2. $3\sin^2(\alpha) = 2(1 - \cos\alpha)$
   a) Find all the solutions for $\alpha$ in the range $-180° \leqslant \alpha \leqslant 360°$. **[7 marks]**
   b) Hence or otherwise, find all the solutions to $3\sin^2(2x) = 2(1 - \cos 2x)$ **[4 marks]**

# Binomial Expansion

Binomial expansion deals with the expansion of a bracket containing a binomial when it is raised to a power: $(ax + b)^n$

## Multiplying Out Brackets

**Example**

$$(ax + b)^3 = (ax + b)(ax + b)(ax + b)$$

$$= (ax + b)((ax)^2 + axb + bax + b^2)$$

$$= (ax + b)\left((ax)^2 + 2abx + b^2\right)$$

$$= (ax)^3 + 2a^2bx^2 + ab^2x + a^2bx^2 + 2ab^2x + b^3$$

$$= (ax)^3 + 3(ax)^2 b + 3(ax)b^2 + b^3$$

By using the patterns that emerge in the coefficients, you can get to this result with simpler working, especially if you only need to find coefficients of particular terms.

## Simple Binomial Expansions

For the simple binomial expansions, $(1 + x)^n$, where $n$ is a positive integer, the coefficients match the numbers found in Pascal's triangle.

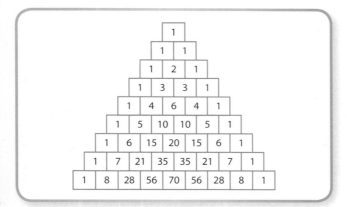

Within Pascal's triangle, each number is calculated by adding the two numbers above it. This can be a quick way to generate the coefficients for relatively low values of $n$. To find the coefficients for an expansion in the form $(ax + b)^n$, use the numbers in the row that starts 1, $n$, … For example, the coefficients to use for $(ax + b)^4$ are 1, 4, 6, 4, 1.

An alternative way is to use **combinations** to generate each coefficient associated with a term. This can be very efficient if you are only interested in particular terms, or where $n$ is a relatively large number.

**Example**

Consider $(1 + x)^4$.

The expansion will be in the form:

$$p(1)^4(x)^0 + q(1)^3(x)^1 + r(1)^2(x^2) + s(1)^1(x)^3 + t(1)^0(x)^4$$

The values of $p$, $q$, $r$, $s$ and $t$ come from the number of ways in which that term can be generated, i.e.

$$(1 + x)^4 = (1 + x)(1 + x)(1 + x)(1 + x)$$

There is one way to get the $x^0$ term from this expansion.

$$(1 + x)(1 + x)(1 + x)(1 + x)$$

There are four ways to generate the $x^1$ term in this expansion.

To calculate the number of combinations that produce each term, $x^a$, we use the theory of combinations. The number of ways each term can be generated is $\dfrac{n!}{r!(n - r)!}$, which can also be written as $(n, r) = {}_nC_r = {}^nC_r = \dbinom{n}{r}$, where $n$ is the power and $r$ is the index of the $x$ term being generated.

**Example**

What is the coefficient of the $x^6$ term in the expansion of $(1 + x)^8$?

$n = 8$ and $r = 6$

The coefficient is $\dfrac{8!}{6!(8 - 6)!} = \dfrac{8 \times 7 \times 6 \times 5 \times 4 \times 3 \times 2 \times 1}{6 \times 5 \times 4 \times 3 \times 2 \times 1 \times (2 \times 1)}$

$$= \dfrac{8 \times 7}{(2 \times 1)} = 28$$

This is the seventh value in the $n = 8$ row of Pascal's triangle.

This can be calculated using the factorial button on a calculator or by using the 'combination key' labelled using one of the representations above, generally ${}_nC_r$ or ${}^nC_r$.

# Complex Binomial Expansions

## $(a + bx)^n$, where $n$ is a Positive Integer

For more complex binomial expansions, $(a+bx)^n$, where $n$ is a positive integer and $a$ and $b$ are constants, the coefficients are the product of $^nC_r$ (the number of combinations creating the term) and the corresponding powers of $a$ and $b$:

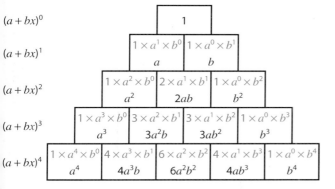

$(a+bx)^0$

$(a+bx)^1$

$(a+bx)^2$

$(a+bx)^3$

$(a+bx)^4$

You can save time if you know that the first two coefficients of any expansion are always 1 and $n$; the symmetry of Pascal's triangle also means, for example, that $^7C_2 = {}^7C_5 = 21$.

### Example
Expand $(2+3x)^5$ fully.

$$(2+3x)^5 = {}^5C_5 \times 2^5 + {}^5C_4 \times 2^4 \times (3x)^1$$
$$+ {}^5C_3 \times 2^3 \times (3x)^2 + {}^5C_2 \times 2^2 \times (3x)^3$$
$$+ {}^5C_1 \times 2^1 \times (3x)^4 + {}^5C_0 \times 2^0 \times (3x)^5$$
$$= 32 + 240x + 720x^2 + 1080x^3 + 810x^4 + 243x^5$$

Questions often ask for an expansion up to a certain term, rather than expecting the full expansion. This means larger powers of $n$ can be used without lots of time being spent crunching numbers. The introduction of algebraic coefficients makes the topic slightly harder. Watch out for negatives as well!

### Example
The coefficient of the $x^3$ term in the expansion of $(a-3x)^8$ is $-48\,384$. Find the value of $a$.

$$^8C_3 \times (-3)^3 \times (a)^5 = -48\,384$$
$$56 \times -27 \times a^5 = -48\,384$$
$$a^5 = \frac{-48\,384}{-1512} = 32$$
$$a = \sqrt[5]{32} = 2$$

The term $bx$ can also be a simple function of $x$ and the expansion will still work. So a further generalisation could be made such that in the expansion of $(a + f(x))^n$ each **term** can be found using $^nC_r a^{n-r} (f(x))^r$. If asked to find a given term $px^q$, you might need to give some thought as to what power, $r$, of $f(x)$ would generate the given term $px^q$.

### Example
What is the coefficient of $x^4$ in the expansion of $(3+2\sqrt{x})^{12}$?

First identify the value of $r$: $x^4 = (x^{\frac{1}{2}})^r = x^{\frac{r}{2}}$

$r = 8$

$$^{12}C_8 \times 3^{12-8} \times (2x^{\frac{1}{2}})^8 = 495 \times 81 \times 256x^4$$
$$= 10\,264\,320x^4$$

The coefficient is $10\,264\,320$.

## $(1 + bx)^n$, where $n$ is not a Positive Integer

If $n$ is not a positive integer, you can no longer use $^nC_r$. The expansion of the binomial is still possible but there will be an infinite number of terms.

$$(1+x)^n = 1 + nbx + \frac{n(n-1)}{2!}(bx)^2 + \frac{n(n-1)(n-2)}{3!}(bx)^3 + \dots$$

The factorial key [!] can be used. This multiplies the number by each integer value below it, i.e. $4! = 4 \times 3 \times 2 \times 1 = 24$.

However, the expansion is only valid when $|bx| < 1$.

### Example
Find the binomial expansion of $(1+4x)^{\frac{1}{2}}$ up to and including the term in $x^4$, and state for which values of $x$ the expansion is valid.

$$(1+4x)^{\frac{1}{2}} = 1 + \frac{1}{2}(4x) + \frac{\left(\frac{1}{2}\right)\left(-\frac{1}{2}\right)}{2!}(4x)^2$$
$$+ \frac{\left(\frac{1}{2}\right)\left(-\frac{1}{2}\right)\left(-\frac{3}{2}\right)}{3!}(4x)^3 + \frac{\left(\frac{1}{2}\right)\left(-\frac{1}{2}\right)\left(-\frac{3}{2}\right)\left(-\frac{5}{2}\right)}{4!}(4x)^4 + \dots$$
$$= 1 + 2x - 2x^2 + 4x^3 - 10x^4 + \dots$$

This expansion is valid for $|4x| < 1$, i.e. $-\frac{1}{4} < x < \frac{1}{4}$.

### $(a + bx)^n$, where $n$ is not a Positive Integer

By considering this to be the expansion of
$\left(a\left(1 + \frac{b}{a}x\right)\right)^n = a^n\left(1 + \frac{b}{a}x\right)^n$, you can adjust the
expansions above to say that:

$(a + bx)^n$

$= a^n\left(1 + n\left(\frac{b}{a}x\right) + \frac{n(n-1)}{2!}\left(\frac{b}{a}x\right)^2 + \frac{n(n-1)(n-2)}{3!}\left(\frac{b}{a}x\right)^3 + \ldots\right)$

$= a^n + a^n n\left(\frac{b}{a}x\right) + a^n\frac{n(n-1)}{2!}\left(\frac{b}{a}x\right)^2 + a^n\frac{n(n-1)(n-2)}{3!}\left(\frac{b}{a}x\right)^3 + \ldots$

For the expansion to be valid, $\left|\frac{b}{a}x\right| < 1$.

---

**Example**

Find the binomial expansion of $\frac{1}{(2 - 3x)^2}$ up to and
including the term in $x^3$, and explain why this
wouldn't be a valid expansion if $x = 0.67$.

$(2 - 3x)^{-2}$

$= 2^{-2}\left(1 - \frac{3}{2}x\right)^{-2}$

$= 2^{-2}\left(1 + (-2)\left(\frac{-3}{2}x\right) + \frac{-2(-3)}{2!}\left(\frac{-3}{2}x\right)^2 + \frac{-2(-3)(-4)}{3!}\left(\frac{-3}{2}x\right)^3 + \ldots\right)$

$= \frac{1}{4}\left(1 + 3x + \frac{27}{4}x^2 + \frac{27}{2}x^3 + \ldots\right)$

$= \frac{1}{4} + \frac{3}{4}x + \frac{27}{16}x^2 + \frac{27}{8}x^3 + \ldots$

The expansion is valid for $\left|\frac{-3}{2}x\right| < 1$, i.e. $-\frac{2}{3} < x < \frac{2}{3}$.

As 0.67 is outside this range, the expansion is
not valid.

---

Sometimes a question will require problem solving
to find $n$, $a$ or $b$. By considering the expansion and
comparing coefficients, you can obtain equations
which you can solve to find the unknown values.

---

**Example**

The first three terms in the expansion of $\sqrt[3]{(a + bx)}$,
where $a$ and $b$ are positive integers, are

$\sqrt[3]{5} + \frac{4\sqrt[3]{5}}{15}x - \frac{16\sqrt[3]{5}}{225}x^2 + \ldots$

**a)** Find the fourth term in the expansion.

$\sqrt[3]{(a + bx)} = (a + bx)^{\frac{1}{3}} = a^{\left(\frac{1}{3}\right)}\left(1 + \frac{b}{a}x\right)^{\frac{1}{3}}$

---

$= a^{\frac{1}{3}}\left(1 + \frac{1}{3}\left(\frac{b}{a}x\right) + \frac{\frac{1}{3}\left(-\frac{2}{3}\right)}{2!}\left(\frac{b}{a}x\right)^2 + \frac{\frac{1}{3}\left(-\frac{2}{3}\right)\left(-\frac{5}{3}\right)}{3!}\left(\frac{b}{a}x\right)^3 + \ldots\right)$

$= a^{\frac{1}{3}} + a^{\frac{1}{3}} \times \frac{1}{3}\left(\frac{b}{a}x\right) + a^{\frac{1}{3}} \times \frac{\frac{1}{3}\left(-\frac{2}{3}\right)}{2!}\left(\frac{b}{a}x\right)^2$

$+ a^{\frac{1}{3}} \times \frac{\frac{1}{3}\left(-\frac{2}{3}\right)\left(-\frac{5}{3}\right)}{3!}\left(\frac{b}{a}x\right)^3$

By comparing coefficients:

$\sqrt[3]{5} = a^{\frac{1}{3}}$ $\quad\quad a = 5$

$\frac{4\sqrt[3]{5}}{15} = a^{\frac{1}{3}} \times \frac{1}{3} \times \frac{b}{a} = \frac{\sqrt[3]{5} \times b}{3 \times 5} \rightarrow b = 4$

$\sqrt[3]{5} \times \frac{\frac{1}{3}\left(-\frac{2}{3}\right)\left(-\frac{5}{3}\right)}{3!}\left(\frac{4}{5}x\right)^3 = \frac{64\sqrt[3]{5}}{2025}x^3$

**b)** Hence state the range of values of $x$ for which
the expansion is valid.

The expansion is valid for $-\frac{5}{4} < x < \frac{5}{4}$.

---

### Using Binomial Expansions to Find or Estimate a Value

You can use binomial expansion to **calculate** values.

---

**Example**

Use the binomial expansion of $(2 - 3x)^3$ to find
the value of $1.7^3$ to 3 decimal places. You must
justify your answer.

$(2 - 3x)^3 = 8 - 36x^1 + 54x^2 - 27x^3$

$2 - 0.3 = 1.7$, so let $x = 0.1$

$1.7^3 = 8 - 3.6 + 0.54 - 0.027 = 4.913$

---

You can also use binomial expansion to **estimate**
values. To do this, the binomial expansion, up to a
suitable term, should be shown clearly and a value of $x$
calculated and substituted into the expansion. Bear in
mind the validity of the expansion and thus the limits
on the possible values of $x$.

## Example

**a)** Show that $\dfrac{2}{\sqrt{5}} = \dfrac{1}{\sqrt{1.25}}$

$$\frac{2}{\sqrt{5}} = \frac{2}{\sqrt{4\sqrt{\frac{5}{4}}}} = \frac{2}{2\sqrt{1.25}} = \frac{1}{\sqrt{1.25}}$$

**b)** Use the binomial expansion of $(1+2x)^{-\frac{1}{2}}$ to estimate the value of $\dfrac{2}{\sqrt{5}}$ to 2 decimal places.

$$(1+2x)^{-\frac{1}{2}} = 1 + \left(-\frac{1}{2}\right)(2x) + \frac{\left(-\frac{1}{2}\right)\left(-\frac{3}{2}\right)}{2!}(2x)^2$$

$$+ \frac{\left(-\frac{1}{2}\right)\left(-\frac{3}{2}\right)\left(-\frac{5}{2}\right)}{3!}(2x)^3 + \dots$$

$$= 1 - x + \frac{3}{2}x^2 - \frac{5}{2}x^3 + \dots$$

Now fit the calculation to the required form:

$\dfrac{2}{\sqrt{5}} = (1+2x)^{-\frac{1}{2}}$, bearing in mind that $-\dfrac{1}{2} < x < \dfrac{1}{2}$

for the above expansion to be valid.

$\dfrac{1}{\sqrt{(1+2x)}} = \dfrac{1}{\sqrt{1.25}}$, part **a)** has already provided

the first step

$$1 + 2x = 1.25 \quad \rightarrow \quad 2x = \frac{1}{4}$$

Let $x = \dfrac{1}{8}$

Substitute the $x$-value into the expansion (note the use of the approximately equals sign and the exclusion of the $+\dots$):

$$\frac{2}{\sqrt{5}} \approx 1 - \left(\frac{1}{8}\right) + \frac{3}{2}\left(\frac{1}{8}\right)^2 - \frac{5}{2}\left(\frac{1}{8}\right)^3$$

$$\approx 1 - 0.125 + 0.0234375 - 0.0048828125$$

$$\approx 0.89$$

Note: This question didn't specify the number of terms for the expansion. In general, expect three or four terms of the expansion to be relevant but check that the final term found is small enough to give the accuracy required.

### Links to Other Concepts
● Creating expressions that can then be differentiated or integrated ● Partial fractions
● Probability, permutations and combinations
● Modulus function and inequalities
● Manipulating and simplifying algebraic expressions

## SUMMARY

● A binomial of the form $(1+x)^n$ will have coefficients for its terms that match the numbers in the corresponding row from Pascal's triangle. These values can be calculated using $^nC_r$.

● $(a+b)^n = a^n + {^nC_1}a^{n-1}b^1 + {^nC_2}a^{n-2}b^2$

$$+ \dots + {^nC_r}a^{n-r}b^r + \dots + b^n$$

for $n \in \mathbb{N}$ or $n \in \mathbb{Z}^+$ where $^nC_r = \binom{n}{r} = \frac{n!}{r!(n-r)!}$

● In the expansion of $(a+b)^n$, $n \in \mathbb{N}$ or $n \in \mathbb{Z}^+$, each term is $^nC_r \times a^r \times b^{n-r}$. Note the power of $a$ and the power of $b$ add up to $n$.

● In the expansion of $(a+\mathrm{f}(x))^n$, $n \in \mathbb{N}$ or $n \in \mathbb{Z}^+$, each term can be found using $^nC_r a^{n-r}(\mathrm{f}(x))^r$.

● If $n \notin \mathbb{N}$ or $n \notin \mathbb{Z}^+$, the expansion is an infinite series:

$$(a+bx)^n = a^n + a^n n\left(\frac{b}{a}x\right) + a^n \frac{n(n-1)}{2!}\left(\frac{b}{a}x\right)^2$$

$$+ a^n \frac{n(n-1)(n-2)}{3!}\left(\frac{b}{a}x\right)^3 + \dots$$

For the expansion to be valid, $\left|\frac{b}{a}x\right| < 1$.

● In $(a+b)^n$, the letters $a$ and $b$ can represent constants but also simple functions.

## QUICK TEST

1.  Sara intends to use the binomial expansion of $(4+x)^{-3}$ to estimate the value of $(2.3)^{-3}$. Is this a suitable method to use? Justify your answer.

## PRACTICE QUESTIONS

1.  $(2+bx)^p = \sqrt[3]{128} + 14\sqrt[3]{2}x + \frac{7\sqrt[3]{2}}{2}x^2 + \dots$

    **a)** What is the value of $p$? **[2 marks]**

    **b)** What is the value of $b$? **[2 marks]**

    **c)** What are the possible values of $x$ for which the expansion is valid? **[1 mark]**

    **d)** Use the binomial expansion to estimate the value of $2.06^p$ to a suitable degree of accuracy. **[5 marks]**

# Sequences and Series

Each number in a sequence is called a term and can be expressed as $u_1, u_2, u_3, u_4, u_5, \ldots, u_n$ where $u_2$ is the second term in the sequence. There are different ways to express the rule that links the terms.

## Term-to-Term Rule

You can apply the term-to-term rule to get from one term to the next in a sequence. For example, the term-to-term rule for the sequence 3, 5, 7, 9, ... is "add 2".

A recurrence relation can also be used to express the rule. For example, the recurrence relation for the sequence 3, 5, 7, 9, ... would be $u_{n+1} = u_n + 2$. This means any term ($u_{n+1}$) is 2 more than the previous term ($u_n$).

Given a term-to-term rule and a starting point, you can generate terms of a sequence.

> **Example**
> Given that the fifth term in a sequence is 21 and that the recurrence relation between terms is $u_{n+1} = 2u_n - 1$, write down the first six terms of the sequence.
>
> To find the sixth term, substitute into
> $u_6 = 2u_5 - 1 = 2 \times 21 - 1 = 41$
>
> To find the previous terms, rearrange the recurrence relation: $u_{n+1} + 1 = 2u_n$
>
> $$u_n = \frac{u_{n+1} + 1}{2}$$
>
> Then generate the first four terms of the sequence:
>
> $$u_4 = \frac{u_5 + 1}{2} = \frac{21 + 1}{2} = 11$$
> $$u_3 = \frac{u_4 + 1}{2} = \frac{11 + 1}{2} = 6$$
> $$u_2 = \frac{u_3 + 1}{2} = \frac{6 + 1}{2} = 3.5$$
> $$u_1 = \frac{u_2 + 1}{2} = \frac{3.5 + 1}{2} = 2.25$$
>
> The first six terms are 2.25, 3.5, 6, 11, 21, 41.

## $n$th Term Rule

While term-to-term rules are useful in finding small runs within a sequence, it would be time consuming to use them to find, for example, the 70th term in a sequence having been given only the first term. For this reason, the $n$th term rule defines a term $u_n$ (the $n$th term) in relation to $n$ (its term number).

> **Example**
> The $n$th term of a sequence is defined as $u_n = 3n + 2$. Find the 50th term.
>
> $u_{50} = 3 \times 50 + 2 = 152$

## Increasing, Decreasing and Periodic Sequences

An **increasing sequence** is one where $u_{n+1} > u_n$. Put another way, it is a sequence where the terms increase as the term number increases.

> **Example**
> $u_{n+1} = 2u_n + 1$ with $u_1 = 3$
>
> This gives a sequence 3, 7, 15, 31, ...

A **decreasing sequence** is one where $u_{n+1} < u_n$. Put another way, it is a sequence where the terms decrease as the term number increases.

> **Example**
> $u_{n+1} = 2u_n + 1$ with $u_1 = -2$
>
> This gives a sequence −2, −3, −5, −9, ...

A **periodic sequence** repeats over an interval. $u_{n+a} = u_n$. The period of the sequence is the number of terms over which the sequence is repeated.

> **Example**
> $u_n = \cos\left(\frac{n\pi}{4}\right) + 1$
>
> This gives a sequence: $\frac{2 + \sqrt{2}}{2}$, 1, $\frac{2 - \sqrt{2}}{2}$, 0, $\frac{2 - \sqrt{2}}{2}$, 1,
> $\frac{2 + \sqrt{2}}{2}$, 2, $\frac{2 + \sqrt{2}}{2}$, 1, $\frac{2 - \sqrt{2}}{2}$, 0, $\frac{2 - \sqrt{2}}{2}$, 1, $\frac{2 + \sqrt{2}}{2}$, ...
>
> This sequence has a period of 8.

# Limit of a Sequence

Some sequences tend towards a certain value; in these cases there is a limit to the sequence. The limit of the sequence, as $n \to \infty$, is considered to occur at the point where $L = u_n = u_{n+1}$

## Example

A sequence is defined by the recurrence relationship $u_{n+1} = \frac{1}{4}u_n + 2$, where $u_1 = 2$.

Find the value of $L$, the limit of the sequence.

The sequence is $2, \frac{5}{2}, \frac{21}{8}, \frac{85}{32}, \ldots$

$$L = \frac{1}{4}L + 2$$

$$\frac{3}{4}L = 2$$

$$L = \frac{2}{\left(\frac{3}{4}\right)} = \frac{8}{3}$$

So as $n \to \infty$, $u_n \to \frac{8}{3}$

To check the above calculation, you can put the values into a calculator. By keying in the first term (2), then pressing equals, you can get the calculator to do the recurrence by using the ANS key as $u_n$ in the calculation $\frac{1}{4}u_n + 2$.

This can be used to generate a number of terms in a sequence defined by a recurrence relationship and, by pressing the equals button lots of times, it will start to show what the limit would be. In this case, nine presses gives 2.66666412... and 11 presses gives 2.66666650...

Since $\frac{8}{3} = 2.6666666666\ldots$ this supports the answer above and is a relatively quick way to check that no mistakes have been made.

## Arithmetic Sequences

An arithmetic sequence has a common difference between each term. The term-to-term rule would be "add $d$", where $d$ is a constant.

The generalised form of an arithmetic sequence is:

$a, a+d, a+2d, a+3d, a+4d, \ldots, a+(n-1)d, \ldots$

The $n$th term for an arithmetic sequence is $a + (n - 1)d$ where $a$ is the first term, $n$ is the term number and $d$ is the common difference.

## Example

An arithmetic sequence has the first four terms 24, 21, 18, 15, ... Find an expression for the $n$th term and hence find the 10th term.

The first term is $a = 24$
The common difference is $d = -3$

$u_n = 24 - 3(n - 1)$, $u_{10} = -3$

More complex questions may not give the first term of the sequence, may not directly show the common difference, or may include algebraic terms.

## Example

The third term in an arithmetic sequence is $2p + 4$, the fifth term is $3p$, and the ninth term is 37.

Find an expression for the $n$th term of the sequence.

Start by considering what is known:

$$a + 2d = 2p + 4 \qquad \{1\}$$
$$a + 4d = 3p \qquad \{2\}$$
$$a + 8d = 37 \qquad \{3\}$$

This forms a set of simultaneous equations to solve.

$\{2\} - \{1\}$ gives $2d = p - 4$

$$p = 2d + 4$$

Substitute into $\{1\}$: $a + 2d = 2(2d + 4) + 4 = 4d + 12$

$$a = 2d + 12$$

Substitute into $\{3\}$: $\qquad 2d + 12 + 8d = 37$

$$10d = 37 - 12 = 25$$

$$d = 2.5$$

Rearrange $\{3\}$ and substitute in to find $a$:

$$a = 37 - 8d = 37 - 8 \times 2.5$$

$$a = 17$$

$n$th term of an arithmetic sequence is $a + (n - 1)d$:

$$u_n = 17 + (n - 1) \times 2.5 = 2.5n + 14.5$$

## Geometric Sequences

A geometric sequence has a common ratio between each term. The term-to-term rule would be 'multiply by $r$', where $r$ is a constant.

The generalised form of a geometric sequence is:

$$a, \quad ar, \quad ar^2, \quad ar^3, \quad ar^4, \quad ..., \quad ar^{(n-1)}, \quad ...$$

The $n$th term for a geometric sequence is $ar^{(n-1)}$ where $a$ is the first term, $n$ is the term number and $r$ is the common ratio.

> **Example**
>
> Find an expression for the $n$th term in the sequence 24, 12, 6, 3, ... and hence find the 10th term.
>
> The first term is $a = 24$
>
> The common ratio is $r = \dfrac{1}{2}$
>
> The $n$th term is $u_n = 24 \times \left(\dfrac{1}{2}\right)^{n-1}$
>
> The 10th term is $u_{10} = 24 \times \left(\dfrac{1}{2}\right)^{10-1} = 24 \times \left(\dfrac{1}{2}\right)^9 = \dfrac{3}{64}$

Questions can be made more difficult by not giving the first term of the sequence, by not directly showing the common difference, or by including algebraic terms. Again, by setting out the terms and setting up simultaneous equations, you can find the values needed.

## Series and Summing

A series is the sum of a sequence of numbers.

> **Example**
>
> $2 + 4 + 6 + 8 + ...$ is an arithmetic series.
>
> $80 + 20 + 5 + 1.25 + ...$ is a geometric series.

$S_n$ refers to the sum of the first $n$ terms of a series.

$\Sigma(u_n)$ sigma notation is also used to denote the sum of a sequence.

You can use subscript to define a sum that is slightly more complicated:

- ● $\displaystyle\sum_{n=2}^{5}(u_n)$ means the sum of the terms between the second and the fifth term inclusive.

- ● $S_p = \displaystyle\sum_{n=1}^{p}(u_n)$

> **Example**
>
> $$\sum_{n=2}^{5}(n^2 - 3) = (2^2 - 3) + (3^2 - 3) + (4^2 - 3) + (5^2 - 3)$$
>
> $$= 1 + 6 + 13 + 22 = 42$$

## Sum of an Arithmetic Series

$S_n = \dfrac{n}{2}(a + l)$, where $a$ is the first term and $l$ is the last term.

$S_n = \dfrac{n}{2}(2a + (n-1)d)$, where $a$ is the first term and $d$ is the common difference.

You may be asked to prove the summation formulae.

> **Example**
>
> Prove that the sum of the first $n$ terms of an arithmetic sequence with common difference $d$ is $S_n = \dfrac{n}{2}(2a + (n-1)d)$.
>
> The sum to $n$ terms can be written as a series from $a$ (the first term) to $a + (n-1)d$ (the $n$th term). By considering the sequence both forwards and backwards, it is possible to see that the sum of each pair of terms is the same:
>
> $S_n = a + (a + d) + (a + 2d) + ... + (a + (n-3)d) + (a + (n-2)d) + (a + (n-1)d)$
>
> $S_n = (a + (n-1)d) + (a + (n-2)d) + (a + (n-3)d) + ... + (a + 2d) + (a + d) + a$
>
> $2S_n = (2a + (n-1)d) + (2a + (n-1)d) + (2a + (n-1)d)$
> $\quad + ... + (2a + (n-1)d) + (2a + (n-1)d) + (2a + (n-1)d)$
> $\quad = n(2a + (n-1)d)$
>
> $S_n = \dfrac{n}{2}(2a + (n-1)d)$

Exam questions will include problem-solving elements and are unlikely to give all the key information easily. Try to find a way of working out any missing information (this will often include simultaneous equations).

> **Example**
>
> The 19th term of an arithmetic sequence is 30 and $S_{44} = 1551$. Find the values of the first term, $a$, and the common difference, $d$.
>
> $u_{19} = a + 18d = 30$
>
> $S_{44} = 22(2a + 43d) = 1551$

Solving the simultaneous equations:

$$a = 30 - 18d$$

$$1551 = 22(2(30 - 18d) + 43d)$$

$$1551 = 1320 - 792d + 946d$$

$$154d = 1551 - 1320 = 231$$

$$d = \frac{231}{154} = 1.5$$

$$a = 30 - 18 \times 1.5 = 3$$

In some cases the number of terms are few enough that each term can be found easily, and thus the sum. In other cases, finding $\sum_{n=a} (u_n)$ will require the use of the summation formula:

$$\sum_{n=a}^{b} (u_n) = \sum_{n=1}^{b} (u_n) - \sum_{n=1}^{a-1} (u_n) = S_b - S_{(a-1)}$$

Note that the sum to $(a - 1)$ is subtracted from the sum to $b$ since the term $u_a$ is included in $\sum_{n=a}^{b} (u_n)$.

**Example**

Find the value of $\sum_{n=10}^{15} u_n$

$a = 3$ and $d = 1.5$

$$\sum_{n=10}^{15} u_n = S_{15} - S_9$$

$$= \frac{15}{2}(2 \times 3 + 14 \times 1.5) - \frac{9}{2}(2 \times 3 + 8 \times 1.5)$$

$$= 202.5 - 81 = 121.5$$

## Sum of a Geometric Series

$S_n = \frac{a(1 - r^n)}{1 - r} = \frac{a(r^n - 1)}{r - 1}$, where $a$ is the first term and $r$ is the common ratio. You may need to prove summation formulae.

**Example**

Prove that the sum of the first $n$ terms of a geometric sequence with common ratio $r$ is $S_n = \frac{a(1 - r^n)}{1 - r}$.

$$S_n = a + ar + ar^2 + ar^3 + \ldots + ar^{n-3} + ar^{n-2} + ar^{n-1}$$

$$rS_n = ar + ar^2 + ar^3 + \ldots + ar^{n-3} + ar^{n-2} + ar^{n-1} + ar^n$$

$$S_n - rS_n = a - ar^n$$

$$S_n(1 - r) = a(1 - r^n)$$

$$S_n = \frac{a(1 - r^n)}{1 - r}$$

### Summing Geometric Series to Infinity

For convergent geometric series: $S_\infty = \frac{a}{1 - r}, \; |r| < 1$

A convergent geometric series has individual terms that get closer and closer in value to 0 as $n$ increases; this happens when the modulus of the common ratio is less than 1 ($|r| < 1$). The sum of the terms will converge, or tend towards, a certain value. By considering the formula for $S_n = \frac{a(1 - r^n)}{1 - r}$, you can see that as $n \to \infty$, $(1 - r^n) \to 1$ and so $S_n \to \frac{a}{1 - r}, \; |r| < 1$.

**Example**

The $n$th term of a geometric series is $81 \times r^{n-1}$. The fourth term is 3. Find the sum to infinity for this series.

$$u_n = ar^{(n-1)} = 81 \times r^{n-1}$$

$$a = 81$$

$$u_4 = ar^3$$

$$81r^3 = 3$$

$$r^3 = \frac{3}{81}$$

$$r = \sqrt[3]{\frac{3}{81}} = \sqrt[3]{\frac{1}{27}} = \frac{1}{3}$$

As $|r| < 1$, $S_\infty = \frac{a}{1 - r}$

$$S_\infty = \frac{81}{1 - \frac{1}{3}} = \frac{243}{2} = 121.5$$

You may need to problem solve and use algebraic terms to reach the final answer. Key information to consider is the values of $a$ and $r$ ($a$ and $d$ for an arithmetic series). If given two terms, you can divide them in order to find $r$ and hence $a$.

### Example

$$\sum_{n=7}^{15} u_n = 5472, u_6 = -64p, u_{11} = 2048p.$$

Find the value of $p$.

$$u_6 = ar^5 = -64p \text{ and } u_{11} = ar^{10} = 2048p$$

$$\frac{ar^{10}}{ar^5} = \frac{2048p}{-64p}$$

$$r^5 = -32$$

$$r = \sqrt[5]{-32} = -2$$

$$-32a = -64p$$

$$a = 2p$$

$$\sum_{n=7}^{15} u_n = S_{15} - S_6 = \frac{a(1-r^{15})}{1-r} - \frac{a(1-r^6)}{1-r}$$

$$= \frac{a(r^6 - r^{15})}{1-r}$$

$$5472 = \frac{2p((-2)^6 - (-2)^{15})}{1--2}$$

$$5472 = \frac{2p(32\,832)}{3} = 21\,888p$$

$$p = \frac{5472}{21\,888} = \frac{1}{4}$$

## Context-Based Questions

### Example

A cylindrical garden water butt leaks at a rate proportional to the height of the water contained within it. $\frac{1}{8}$ of the water leaks each day and the gardener tops up the butt by 20 cm each day. On Monday evening the gardener tops the water level up to 180 cm. The gardener notices that after a long time the water level seems to be remaining constant. What is the water level tending towards?

It is always advisable to try to work through a few 'days' to set up the sequence/series:

On day 1 (Mon) there is 180 cm of water in the water butt.

On day 2 (Tues) there is $180 \times \frac{7}{8} + 20 = 177.5$ cm

On day 3 (Wed) there is $177.5 \times \frac{7}{8} + 20 = 175.3125$ cm

$$u_{n+1} = \frac{7}{8}u_n + 20$$

By finding the term-to-term rule, you can consider the limit of the sequence:

Using the limit $L = \frac{7}{8}L + 20$

$$\frac{1}{8}L = 20 \qquad \therefore \qquad L = 160 \text{ cm}$$

## SUMMARY

- **Term-to-term rules define one term in relation to a previous term.**

  **Generally $u_{n+1} = f(u_n)$.**
  - To find subsequent and previous terms, at least one term must be given.
  - If $u_n$ is given, the inverse function is needed to decrease the term number.

- **The $n$th term defines a term in relation to its position in a sequence or series, i.e. $u_n = f(n)$. Substituting a term number into the function will find the value of that term.**

- **Sequences and series can be defined as increasing, decreasing or periodic.**

- **The limit of a sequence is the value of $u_n$ as $n \to \infty$. It can be found by considering the point where $L = u_n = u_{n+1}$**

- **Arithmetic sequences and series:**

  $a, a+d, a+2d, a+3d, a+4d, \ldots, a+(n-1)d, \ldots$

  **The $n$th term for an arithmetic sequence is $a+(n-1)d$.**

  $$S_n = \frac{n}{2}(2a+(n-1)d)$$

  $$S_n = \frac{n}{2}(a+l)$$

- **Geometric sequences and series:**

  $a, ar, ar^2, ar^3, ar^4, \ldots, ar^{(n-1)}, \ldots$

  **The $n$th term for a geometric sequence is $ar^{(n-1)}$.**

  $$S_n = \frac{a(1-r^n)}{1-r} = \frac{a(r^n-1)}{r-1}$$

  $$S_\infty = \frac{a}{1-r}, |r| < 1$$

- $$\sum_{n=a}^{b}(u_n) = \sum_{n=1}^{b}(u_n) - \sum_{n=1}^{a-1}(u_n) = S_b - S_{(a-1)}$$

**Links to Other Concepts**

● Algebraic manipulation   ● Number theory and sets   ● Factorisation   ● Indices and logs
● Problem solving   ● Polynomials   ● Quadratics   ● Inverse functions

## QUICK TEST

1. For each sequence, find the missing term(s) and determine if it is:

   i) geometric, arithmetic, neither or impossible to tell

   ii) increasing, decreasing, periodic, none or impossible to tell.

   a) 0.2, 0.6, 1.8, 5.4, 16.2, ◆, 145.8, …

   b) $\frac{3}{4}, \frac{5}{4}, ◆, \frac{1}{4}, \frac{3}{4}, \frac{5}{4}, …$

   c) 32 768, 8192, ◆, 512, 128, 32, 8, ◆, …

   d) $u_{n+1} = 0.2u_n + 1$, $u_2 = 4.8$
      $u_1 = ◆$, $u_8 = ◆$, $L = ◆$

2. The Fibonacci sequence is defined by the term-to-term relationship $u_{n+2} = u_n + u_{n+1}$, $u_1 = 1$, $u_2 = 1$. Find all the terms up to and including $u_7$.

3. Find the term-to-term rule, the $n$th term expression and hence the specified term (in brackets) for the following sequences:

   a) 15, 12, 9, 6, …           $(u_{100})$

   b) 0.2, 0.8, 3.2, 12.8, …      $(u_8)$

4. In a geometric sequence, the fifth term is 16 and the second term is 2000.

   a) Find the common ratio.

   b) Find the first term.

   c) What is $\sum_{n=6}^{10}(u_n)$?

5. An arithmetic sequence has a fourth term of 72 and a sixth term of 60. What is the fifth term?

6. $u_{n+1} = 0.3u_n + a$, $u_1 = a$

   a) Find an expression for $L$ in terms of the first term, $a$.

   b) What value of $a$ would give $L = 2$?

## PRACTICE QUESTIONS

1. Emily has written down an arithmetic sequence but her cat has walked over the desk, leaving footprints in the wet ink:

| Term number | $u_1$ | $u_2$ | $u_3$ | $u_4$ | $u_5$ | $u_6$ | $u_7$ |
|---|---|---|---|---|---|---|---|
| Term |  |  | $3q$ |  | $q^2 - 2$ |  |  |

   a) Show that $2d = q^2 - 3q - 2$.   **[3 marks]**

   b) Find an expression for $a$ in terms of $q$. **[2 marks]**

   c) Given that $S_5 = 75$, find the value of $q$.   **[4 marks]**

2. The difference between the terms of a geometric series $(u_n)$ and an arithmetic series $(v_n)$ forms a new series $(w_n = u_n - v_n)$:

   $3193.6 + 1593.5 + 793.4 + 393.3 + …$

   a) Given that $u_1 = 500v_1$, find the values of $u_1$ and $v_1$.   **[4 marks]**

   The geometric series is convergent and

   $$\sum_{n=2}^{\infty}(u_n) = 3200.$$

   b) Find the value of the common ratio in the series $u_n$.   **[4 marks]**

   c) Hence or otherwise, find the value of $\sum_{n=1}^{10}(w_n)$.   **[7 marks]**

# Exponentials

## General Exponential Functions

**Exponential** comes from the word 'exponent', meaning the power or **index**. It is when $f(x) = a^x$, $a > 0$. It is defined for positive values of $a$ only, as a **negative** number raised to a power creates a **discontinuous** set of results. If $0 < a < 1$, the function can be rewritten with a negative index. For example, the function $f(x) = \left(\frac{1}{3}\right)^x$ can be rewritten as $f(x) = 3^{-x}$.

When $a > 0$, the resulting graph is a continuous curve that tends towards zero as $x$ tends to negative infinity. As $x$ increases, so does the $y$-value. The rate at which the $y$-value is increasing (i.e. the gradient of the graph) is also increasing. All graphs in the form $y = a^{kx}$ will have a $y$-intercept of 1, as anything raised to the power of 0 is 1.

By changing the value of $a$, the gradient of the graph changes. A larger value of $a$ will give a steeper curve.

### Transformations with the General Graph

Consider the graph $y = f(x) = p^x$.

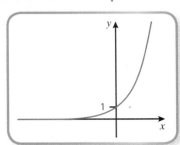

### $y = f(ax) = p^{ax}$, where $a$ is a positive constant

The graph is stretched horizontally by $\frac{1}{a}$. All $y$-intercepts stay in the same place. Each $x$-coordinate is $\times \frac{1}{a}$.

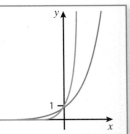

### $y = f(x + a) = p^{x+a}$

The graph is moved left by $a$, or translated by $\begin{pmatrix} -a \\ 0 \end{pmatrix}$. If $a$ is negative, the movement will be to the right.

### $y = -f(x) = -p^x$

The graph is reflected in the $x$-axis.

### $y = f(x) + a = p^x + a$

The graph is moved up by $a$, or translated by $\begin{pmatrix} 0 \\ a \end{pmatrix}$. If $a$ is negative, the movement will be downwards.

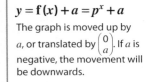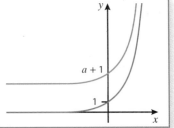

### $y = f(-x) = p^{-x} = \left(\frac{1}{p}\right)^x$

The graph is reflected in the $y$-axis.

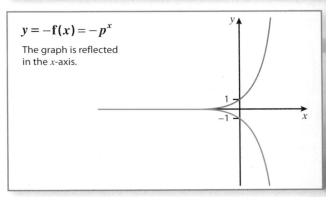

### $y = af(x) = ap^x$, where $a$ is a positive constant

The graph is stretched vertically by $a$. All $x$-intercepts stay in the same place. Each $y$-coordinate is $\times a$.

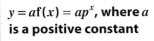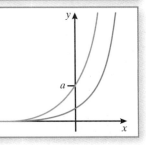

The exponential graphs are horizontal translations and vertical stretches of each other. This is because a horizontal translation of $\begin{pmatrix} -b \\ 0 \end{pmatrix}$ is created by $f(x + b) \rightarrow a^{x+b} = a^b \times a^x$. This is the same as $a^b f(x)$, which is a vertical stretch by a factor of $a^b$.

## Example

The function $f(x) = 3^x$ has a translation applied to it of $\begin{pmatrix} 3 \\ 0 \end{pmatrix}$. What will be the value of the $y$-intercept for the translated graph? Describe the transformation as a stretch.

$$f(x - 3) = 3^{x-3} = 3^{-3} \times 3^x = \frac{1}{27} f(x)$$

The new $y$-intercept will be $3^{-3} = \frac{1}{27}$.

This is seen from either it being the stretch factor in the $y$-direction or by substituting $x = 0$ into $f(x - 3)$.

The transformation could be a stretch in the $y$-direction with a scale factor of $\frac{1}{27}$.

## The Exponential Functions $y = e^x$

e (2.71828182…) is a constant, irrational number. e is the value for which the gradient of the function is equal to the function. There is an $e^{\blacksquare}$ button on all scientific and graphical calculators.

## Example

The gradient of the graph $y = e^x$ at $x = 2$ is $e^2 = 7.389056\ldots$

Sometimes the question will specify a form for the final answer. At other times, the final form will be based on common sense and context. The important thing is to keep it accurate until the final answer.

More generally, **the gradient of the graph $y = ae^{kx}$ at any point is $ake^{kx}$.**

Exponential growth is used to model various real-life problems, from spread of illness to the half-life of radioactive material and compound interest. The exponential function $y = e^x$ is a key part in modelling such situations.

### Links to Other Concepts

- Graph transformations ● Coordinate geometry ● Rates of growth and decay – modelling ● Powers and indices
- Logarithms ● Solving equations

---

### QUICK TEST

1. $y = 3e^{(x-1)}$

   a) What is the exact value of the $y$-intercept?

   b) What is the $x$-value when $y = 3$?

2. A population is described as $P = ae^{-0.1t}$. At $t = 0$, $P = 53\,000$.

   a) What is the value of $a$?

   b) What is the value of $P$ when $t = 100$?

## PRACTICE QUESTIONS

1. Sound pressure ($P$) is measured in dBspl. $P$ is related to the distance ($d$) from the source of the sound by the equation $d = a \times 10^{\left(6.5 - \frac{P}{20}\right)}$, where $a$ is a constant.

   a) Sketch the graph of this relationship, plotting $d$ on the $y$-axis and marking clearly any axis-intercepts. **[2 marks]**

   b) Given that $P = 130$ dBspl at $d = 0.5$ m, find the value of $a$. **[3 marks]**

   c) John has to work near the sound source for a maximum of 15 minutes. For legal safe working, $P$ must not exceed 100 dBspl. Find the minimum distance from the source at which it would be safe for John to work. **[2 marks]**

# Logarithms

Logarithms are the inverse function of exponentials. They can be used to find the value of $c$ in the equation $a = b^c$.

This graph shows the function $y = a^x$, where $a > 1$, and its inverse $y = \log_a x$ (said 'log to the base $a$ of $x$').

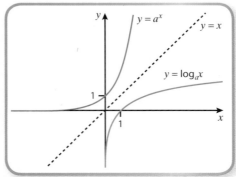

At A-level, logs are defined for $a > 0$ and $x \geq 0$. If $a = 1$, the graph produced would be a horizontal line.

## Special Logarithms

$y = \ln x$ is the shorthand way of writing $y = \log_e x$. This is known as the natural logarithm.

$y = \log x$, where no base is specified, is $y = \log_{10} x$.

## Solving and Rearranging Exponential and Logarithmic Equations

### Example

Find $x$ when $\dfrac{2^{x+3} + 1}{3} = 2731$.

Step 1: Multiply both sides by 3.

$\dfrac{2^{x+3} + 1}{3} \times 3 = 2731 \times 3$

$2^{x+3} + 1 = 8193$

Step 2: Subtract 1 from both sides.

$2^{x+3} + 1 - 1 = 8193 - 1$

$2^{x+3} = 8192$

Step 3: Take log to the base 2 of both sides. (The $x + 3$ falls within the 'brackets' in BIDMAS.)

$\log_2 2^{x+3} = \log_2 8192$

$x + 3 = 13$

Step 4: Subtract 3.

$x = 10$

In questions that ask for exact values, answers should be left in terms of the logarithm if necessary. If a question is a logarithm with the unknown enclosed, then the inverse operation, the exponential, can be used.

### Example

Find the exact value of $t$ when $3 \ln t + 2 = 47$.

Subtract 2 from each side:     $3 \ln t = 45$

Divide each side by 3:     $\ln t = \dfrac{45}{3} = 15$

Having isolated the logarithm, exponentials are used on both sides. In this case the base is e, as it is the 'natural logarithm' ln:

$e^{\ln t} = e^{15}$

As $e^{\ln t} = t$, $t = e^{15}$

## Log Rules

If the base is the same for two, or more, logarithmic terms, they can be combined into a single term. There are three key rules for manipulating logarithms.

Associating them with the index laws can help to understand them, as shown below.

$\log_c a + \log_c b = \log_c ab$

Let $x = \log_c a$ and $y = \log_c b$.

Write in index form:     $a = c^x$ and $b = c^y$

$c^x \times c^y = a \times b$

$c^{x+y} = ab$

Take log base $c$ of both sides:     $\log_c c^{x+y} = \log_c ab$

$x + y = \log_c ab$

Substitute back in for $x$ and $y$:     $\log_c a + \log_c b = \log_c ab$

A similar method of proof can be used to show that

$\log_c a - \log_c b = \log_c \dfrac{a}{b}$    and

$\log_c a^n = n \log_c a$

When using the addition and subtraction rules, there is no coefficient in front of the log. The third rule enables you to convert between a coefficient in front of the logarithm and a power contained within the logarithm, which might be necessary before the addition and subtraction rules can be used.

**Example**

Express $5\log_7 2 + \frac{1}{2}\log_7 8$ as a single logarithm.

$$5\log_7 2 + \frac{1}{2}\log_7 8 = \log_7 2^5 + \log_7 8^{\frac{1}{2}}$$
$$= \log_7 32 + \log_7(2\sqrt{2})$$
$$= \log_7(32 \times 2\sqrt{2})$$
$$= \log_7 64\sqrt{2}$$

## Converting the Base of a Logarithm

$$\log_a b = \frac{\log_c b}{\log_c a}$$

**Example**

Write $\log_2 7 + \log_4 3$ as a single logarithm in the form $\log_4 q$.

$$\log_2 7 = \frac{\log_4 7}{\log_4 2}$$

Given that $\log_4 2 = \frac{1}{2}$ (as $4^{\frac{1}{2}} = 2$):

$$\log_2 7 = \frac{\log_4 7}{\frac{1}{2}} = 2\log_4 7 = \log_4 7^2 \quad (\text{as } a\log_b c = \log_b c^a)$$

$$\log_2 7 + \log_4 3 = \log_4 7^2 + \log_4 3$$
$$= \log_4 49 \times 3 = \log_4 147$$

## Solving More Complex Equations

Combining the laws means more complicated equations can be solved. The aim is always to get all the logarithms together, if possible.

**Example**

$y = \ln 12x + \log 10 - \ln 4$.

**a)** Express $y$ in terms of $\ln(f(x))$.

$y = \ln(12x \div 4) + 1$
$y = \ln(3x) + \ln e$
$y = \ln(3ex)$

**b)** Hence express $x$ in terms of $y$.

Take exponentials of each side: $e^y = e^{\ln(3ex)}$

Using the law $e^{\ln a} = a$: $\qquad e^y = 3ex$

Rearrange to isolate $x$: $\qquad x = \frac{e^y}{3e}$

## Transformations with the General Graph

The graph transformations can be applied to logarithms where $f(x) = \log_p x$.

**Example**

$f(x) = \log_3(x^4 + 4x^3 + 6x^2 + 4x + 1)$

**a)** Find the values of $a$ and $b$ when $f(x)$ is written in the form $a\log_3(x + b)$.

By inspection $x^4 + 4x^3 + 6x^2 + 4x + 1$ is the binomial expansion of $(x + 1)^4$.

$y = 4\log_3(x + 1) \qquad \rightarrow \qquad a = 4, b = 1$

**b)** Hence describe the transformations that map $y = \log_3(x)$ onto $y = f(x)$.

A vertical stretch with scale factor 4 and a translation of $\begin{pmatrix} -1 \\ 0 \end{pmatrix}$.

## Estimating Parameters

By using logarithms it is possible to view the relationships $y = ax^n$ and $y = kb^x$ as linear and hence find unknown values. You can plot the relationship between $x$ and $y$ using a logarithmic scale. It is also possible to rewrite the relationship in terms of a logarithm to give it in the form of a straight line: $y = mx + c$.

$y = ax^n$

Let $Y = \log y$, $X = \log x$ and $c = \log a \quad \rightarrow \quad Y = nX + c$

This is now an equation in the form $y = mx + c$. As such, you can use the linear relationship and given values to find the unknown values.

**Example**

The relationship between two variables is defined as $y = ax^n$, where $a$ and $n$ are constants. Use the data in the table to find the relationship between $x$ and $y$, having found the constants $a$ and $n$.

| x | 2 | 3 | 4 | 5 | 6 |
|---|---|---|---|---|---|
| y | 52 | 175.5 | 416 | 812.5 | 1404 |

Take logs (to base 10) of both sides: $\log y = \log ax^n$

$\log y = n\log x + \log a$

$Y = nX + c$ is now a linear relationship between $Y$ and $X$. Use $Y = \log y$ and $X = \log x$ to find values for the linear relationship:

| $X = \log x$ | log2 | log3 | log4 | log5 | log6 |
|---|---|---|---|---|---|
| $Y = \log y$ | log52 | log175.5 | log416 | log812.5 | log1404 |

The graph is a straight line with gradient $n$ and $y$-intercept of $c$.

To find the gradient, take two pairs of values from the $X$, $Y$ table (gradient is change in $Y$ divided by change in $X$).

Gradient $n = \dfrac{\log 416 - \log 52}{\log 4 - \log 2} = 3$

$Y = 3X + c$

$\log 52 = 3 \log 2 + c$

$c = \log 52 - 3 \log 2$ — Use the log laws to combine the logs. This could be done on a calculator, intermediate step $c = 0.81291\ldots$

$c = \log 52 - \log 2^3$ — On a calculator, use the memory to maintain the full accuracy of the answer, then use $10^c$ to find $a$.

$= \log \dfrac{52}{8} = \log 6.5$

$c = \log a = \log 6.5 \qquad \rightarrow \qquad a = 6.5$

$y = 6.5x^3$

$y = kb^x$

Let $Y = \log y$, $m = \log b$ and $c = \log k \quad \rightarrow \quad Y = mx + c$

### Example

The relationship between two variables is defined as $y = kb^x$, where $b$ and $k$ are constants. Use the data in the table to find the relationship between $x$ and $y$, having found the constants $b$ and $k$.

| $x$ | 5 | 10 | 15 | 20 |
|---|---|---|---|---|
| $y$ | 48 | 1536 | 49 152 | 1 572 864 |

Taking logs of both sides: $\log y = \log kb^x$

Use laws of logs to separate the unknowns on the right-hand side:

$\log y = \log b^x + \log k = x \log b + \log k$

$Y = mx + c$, where $Y = \log y$, $m = \log b$ and $c = \log k$, giving an updated linear relationship between $x$ and $Y$:

| $x$ | 5 | 10 | 15 | 20 |
|---|---|---|---|---|
| $Y = \log y$ | log48 | log1536 | log49 152 | log1 572 864 |

Gradient $m = \dfrac{\log 1536 - \log 48}{10 - 5}$

$= \dfrac{\log \frac{1536}{48}}{5} = \dfrac{1}{5} \log 32$

$= \log 32^{\frac{1}{5}} = \log 2$

As $m = \log b \qquad b = 2$

$Y = mx + c$

$\log 48 = (\log 2) \times 5 + c$

$c = \log 48 - \log 32$

$c = \log \frac{48}{32} = \log \frac{3}{2} \qquad k = \frac{3}{2} = 1.5$

Therefore, $y = 1.5 \times 2^x$

## Real-life Data

When conducting scientific measurements for real, there is always a margin of error in every reading. For this reason the points won't plot perfectly into a straight line (having used logs to gain the linear relationship) as the previous examples would. Given a set of results, a line of best fit can be a good way to estimate the constants within the relationship.

Growth and decay are modelled using exponentials (and therefore logarithms) in many real-life contexts.

Equations are of the form $y = Ae^{bx} + C$ or $y = Ak^t + C$

These equations don't reduce to a simple linear relationship. However, if enough data is available, unknowns can still be found using simultaneous equations. Usually the information for $t = 0$ is given. Using the fact that anything to the power 0 is 1, this would simplify both equations above to $A + C$.

### Example

The population of a species of beetle inhabiting an island is being monitored. The population, $N$, after $t$ years from the start of the study, is modelled as $N = Ae^{kt} + 1500$. At the start of the study, it was estimated that there were 2000 beetles. After 8 years the population is 1601.

a) Find the value of $A$.

At $t = 0$

$2000 = A \times e^{k \times 0} + 1500$

$2000 - 1500 = A \times 1$

$A = 500$

**b)** What size is the population tending towards?

As the population is declining, $k$ is negative.

As $t \to \infty$, $Ae^{kt} \to 0$ $\therefore N \to 1500$ beetles

**c)** What would have been the size of the population after 4 years?

First find $k$ using the information for 8 years:

$1601 = 500e^{k \times 8} + 1500$

$1601 - 1500 = 500e^{8k}$

$e^{8k} = \dfrac{101}{500}$

$8k = \ln\dfrac{101}{500} = -1.5994875\ldots$

$k = -0.1999359\ldots = -0.200$ (3 s.f.)

Now use values for $A$ and $k$ to find $N$ at 4 years:

$N = 500e^{-0.2 \times 4} + 1500 = 1724.6644\ldots$

After 4 years, the population would have been approximately 1725 beetles.

### Links to Other Concepts
- Exponentials
- Graph transformations
- Coordinate geometry
- Rates of growth and decay – modelling
- Powers and indices
- Quadratics
- Differentiation, integration, differential equations

## QUICK TEST

1. Sketch the graphs of $y = 0.2^x$ and $y = \log_{0.2} x$ on the same pair of axes.

2. **a)** Given that $e^{2q-1} = 21$, find the exact value of $q$ in the form $q = a \ln b + c$.

   **b)** Express $q$ in the form $q = \ln(ae^b)$.

3. Find the missing values or expressions.

   **a)** $\log_{2x} p = 4$ $\rightarrow$ $p = \blacklozenge x^4$

   **b)** $\log 2 + \log 6 = \log \blacklozenge$

   **c)** $\dfrac{\ln \blacklozenge - \ln 2}{\ln e^2} = \ln 5$

## PRACTICE QUESTIONS

1. Find the exact solution to the equation
   $\ln(x^2 + x - 2) - \ln(x + 2) - 3 = 0$ **[4 marks]**

2. $y = kb^x$ is converted into a linear equation $Y = mx + c$, where $Y = \log y$. Find an expression for $m$ and $c$ in terms of $p$ and hence complete the table of values. **[9 marks]**

| $x$ | 2 | 4 | 6 | 8 |
|---|---|---|---|---|
| $y$ | $36p$ | 16 | | |

## SUMMARY

- The graph of $\log_a x$ is the reflection of $y = a^x$ in the line $y = x$.

- $\ln x = \log_e x$
- $\log x = \log_{10} x$

- $a^{\log_a b} = b$ and $e^{\ln b} = b$

- $\log_a(a^b) = b$ and $\ln(e^b) = b$

- $\log_a a = 1$ and $\ln(e) = 1$

- $\log_a 1 = 0$ and $\ln(1) = 0$ (assuming $a \neq 1$)

- Laws for combining logs:

  $\log_c a + \log_c b = \log_c ab$

  $\log_c a - \log_c b = \log_c \dfrac{a}{b}$

  $\log_c a^n = n \log_c a$

- Converting the base: $\log_a b = \dfrac{\log_c b}{\log_c a}$

- If asked for an exact answer, it may be left in terms of the logarithm or the exponential.

- Log graphs can be used to estimate parameters in relationships in the form $y = ax^n$ and $y = kb^x$ if data is given for $x$ and $y$.

- By converting the equation $y = ax^n$, it is possible to write in the form $Y = nX + c$, where $Y = \log y$, $X = \log x$ and $c = \log a$.

- By converting the equation $y = kb^x$, it is possible to write in the form $Y = mx + c$, where $Y = \log y$, $m = \log b$ and $c = \log k$.

- For exponential growth and decay problems, use given conditions to find the missing information.

- If asked to consider if the model is realistic, consider long-term implications, e.g. very few populations can increase indefinitely.

# Proof

## Key Concepts and Notation

### Set Notation and Types of Number

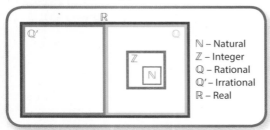

$\mathbb{N}$ – Natural
$\mathbb{Z}$ – Integer
$\mathbb{Q}$ – Rational
$\mathbb{Q}'$ – Irrational
$\mathbb{R}$ – Real

$\mathbb{N}$ – Natural numbers; they are the positive integers and zero. 0, 1, 2, 3, 4, 5, 6, …

$\mathbb{Z}$ – Integers/whole numbers; these include the negatives. …, −4, −3, −2, −1, 0, 1, 2, 3, …

$\mathbb{Q}$ – Rational; these are numbers that can be written accurately as a fraction. They include the integers but also include any fraction or decimal with recurring decimal places.

$\mathbb{Q}'$ – Irrational; numbers that cannot be written accurately as a fraction, e.g. $\sqrt{2}$, $\pi$ and e.

$\mathbb{R}$ – Real numbers; all the numbers mentioned so far. They are the numbers that would fit somewhere on a traditional number line.

The groups can be subdivided into positives and negatives using superscript, e.g. $\mathbb{Z}^-$ would be the negative integers. Just as $\mathbb{Q}'$ is 'not' rational, $\mathbb{N}'$ is 'not' natural.

The symbol $\in$ means 'is an element of'. It is used with a variable and a set of numbers. For example, $x \in \mathbb{N}$ means $x$ belongs to the group of natural numbers. It is an element within this set.

You can use the element symbol with a defined set of numbers, e.g. $n \in \{2, 3, 6, 8, 11\}$ means that $n$ is an element of the set of numbers defined within the brackets.

$\Rightarrow$ means 'implies'.

### Factors, Multiples, Even Numbers, Odd Numbers and Prime Numbers

Factors and multiples deal with natural numbers and the ability to break down numbers and express them as a multiplication of their factors. When working with integers, an even number has a factor of 2. Odd numbers don't have a factor of 2.

A prime number has only two factors: 1 and itself. You must have a method for establishing whether numbers are prime or not (primality).

If presented with a number that is not instantly recognisable as prime or non-prime, you can test primality by dividing the number by the primes up to the square root of the number.

> **Example**
> Determine if 299 and 347 are prime.
>
> By inspection there are no obvious small factors for either value. Quick checks:
>
> 2 – the numbers are not even so 2 is not a factor (even numbers end in a 0, 2, 4, 6 or 8)
>
> 3 – the sum of the digits isn't divisible by 3 so neither number is divisible by 3
>
> 5 – the numbers do not end in a 0 or a 5 so 5 is not a factor
>
> You could carry out these calculations on a calculator. If it gives an integer value, the number is not prime.
>
> $\sqrt{299} = 17.291\ldots$ so try dividing by the primes up to 17:
>
> $299 \div 7 = \frac{299}{7} = 42.7142\ldots$  $\therefore$ 7 is not a factor
>
> $299 \div 11 = \frac{299}{11} = 27.181\ldots$  $\therefore$ 11 is not a factor
>
> $299 \div 13 = 23$  $\therefore$ 13 is a factor, as is 23
>
> $\therefore$ 299 is not prime.
>
> $\sqrt{347} = 18.627\ldots$ so try dividing by the primes up to 18:
>
> $347 \div 7 = \frac{347}{7} = 49.571\ldots$  $\therefore$ 7 is not a factor
>
> $347 \div 11 = \frac{347}{11} = 31.545\ldots$  $\therefore$ 11 is not a factor
>
> $347 \div 13 = \frac{347}{13} = 26.692\ldots$  $\therefore$ 13 is not a factor
>
> $347 \div 17 = \frac{347}{17} = 20.411\ldots$  $\therefore$ 17 is not a factor
>
> $\therefore$ 347 is a prime number.

# Types of Proof

## Proof by Deduction (Direct Proof)

Proof by deduction is based on the idea that from an assumed statement it is possible to deduce a series of other statements until the proof is achieved. Make sure each statement supports the next step.

Deduction often uses a generalisation, e.g. let the number $= n$, where $n$ is any integer value. Then if the rule can be shown to hold true for $n$, it is true for all integer values and so has been proved.

### Example

Here are positive integers organised into a table which is five columns wide. Prove that the sum of any two-by-two square of numbers, taken from within the table, is a multiple of 4.

You can check the interpretation of the question by first considering a specific case:

| 8 | 9 |
|---|---|
| 13 | 14 |

$8 + 9 + 13 + 14 = 44 = 4 \times 11$
$\rightarrow$ this case obeys the rule.

Generalising:
Let $n \in \mathbb{Z}^+$

| $n$ | $n+1$ |
|---|---|
| $n+5$ | $n+6$ |

Within any square, consider $n$ to be the top-left value. Then complete the square in terms of $n$ by using the relative position of the numbers.

Algebraically:
$n + n + 1 + n + 5 + n + 6 = 4n + 12$
$= 4(n + 3)$

Therefore the sum of the numbers within a two-by-two square will always be a multiple of 4.

| 1 | 2 | 3 | 4 | 5 |
|---|---|---|---|---|
| 6 | 7 | 8 | 9 | 10 |
| 11 | 12 | 13 | 14 | 15 |
| 16 | 17 | 18 | 19 | 20 |
| 21 | 22 | 23 | 24 | 25 |
| 26 | 27 | 28 | 29 | 30 |
| 31 | 32 | 33 | 34 | 35 |
| 36 | 37 | 38 | 39 | 40 |
| 41 | 42 | 43 | 44 | 45 |
| 46 | 47 | 48 | 49 | 50 |
| 51 | 52 | 53 | 54 | 55 |
| 56 | 57 | 58 | 59 | 60 |
| 61 | 62 | 63 | 64 | 65 |
| 66 | 67 | 68 | 69 | 70 |
| 71 | 72 | 73 | 74 | 75 |
| 76 | 77 | 78 | 79 | 80 |
| 81 | 82 | 83 | 84 | 85 |
| 86 | 87 | 88 | 89 | 90 |
| 91 | 92 | 93 | 94 | 95 |
| 96 | 97 | 98 | 99 | 100 |
| 101 | 102 | 103 | 104 | 105 |
| 106 | 107 | 108 | 109 | 110 |
| 111 | 112 | 113 | 114 | 115 |
| 116 | 117 | 118 | 119 | 120 |
| 121 | 122 | 123 | 124 | 125 |
| 126 | 127 | 128 | 129 | 130 |

## Proof by Exhaustion

Proof by exhaustion can be used when there are relatively few cases to check. It involves checking the rule with each case and, if they all work, the rule is proved.

### Example

Prove, by exhaustion, that the sum of any pair of different cube numbers less than 100 is not prime.

Cube numbers less than 100: 1, 8, 27, 64

For a number ($n$) to be prime, there has to be one pair of distinct factors (1 and $n$). If a number isn't prime, it will have at least one additional factor pair. Possible pairs:

| | |
|---|---|
| $1 + 8 = 9$ | As $9 = 3 \times 3$, it is not prime. |
| $1 + 27 = 28$ | As $28 = 2 \times 14$, it is not prime. |
| $1 + 64 = 65$ | As $65 = 5 \times 13$, it is not prime. |
| $8 + 27 = 35$ | As $35 = 5 \times 7$, it is not prime. |
| $8 + 64 = 72$ | As $72 = 2 \times 36$, it is not prime. |
| $27 + 64 = 91$ | As $91 = 7 \times 13$, it is not prime. |

Therefore the sum of any pair of different cube numbers less than 100 is not prime.

## Disproof by Counter-Example

Disproof is equally as important as proof. Instead of trying to prove that something is true, you could show that a theory cannot be true by using an example that shows it isn't true for all values.

### Example

Cillian uses the equation $n^2 + 3n + 13$ to generate a sequence of numbers: 17, 23, 31, 41, … He says that all the numbers produced by this sequence will be prime numbers. Prove that Cillian is incorrect.

The sequence so far is all prime numbers. To find a number that isn't prime, it needs to have a common factor that is greater than 1 in each term of the expression. Given that the only factors of 13 are 1 and 13, a counter-example can be found using any multiple of 13, since this will give each term a common multiple of 13.

Let $n = 13$

$$n^2 + 3n + 13 = 13^2 + 3 \times 13 + 13 = 13(13 + 3 + 1)$$
$$= 13 \times 17$$

$\therefore n^2 + 3n + 13$, when $n = 13$, is not prime so Cillian is incorrect.

Any counter-example will do. For the previous example, any multiple of 13 would give at least a second pair of factors, e.g. $n = 26$ would give $n^2 + 3n + 13 = 767 = 13 \times 59$. Remember to complete a proof with a conclusion.

Sometimes disproof would form part of an amendment of the rule, for example by defining the possible set of values more accurately. The previous example shows that the equation will produce primes but not when $n$ is a multiple of 13. This might still be useful to know in some contexts.

### Proof by Contradiction

Proof by contradiction requires an assumption to be set up and a counter to the proof to be made. Then, by taking a series of logical steps, the assumption is disproved and hence the original proof made. Or to put it another way, you assume the statement to be proved is false and then prove that it cannot be the case.

> **Example**
> Prove that there are an infinite number of primes.
>
> To start, make the counter statement: Assume *there are a finite number of primes*
>
> This would mean there is a 'largest prime'. Let the largest prime be $p_z$, the $z$th prime number.
>
> For any number $a$, where $a \in \mathbb{Z}$ and $a > 1$, a multiple of the number could be written as $ab$.
>
> $ab + 1$ is 1 greater than a multiple of $a$ and, as such, is not a multiple of $a$.
>
> Consider multiplying all the primes together up to a certain point:
>
> $p_1 \times p_2 \times p_3 \times p_4 \times p_5 \times \ldots \times p_z = Q$, where $Q$ is a multiple of all the primes up to and including $p_z$.

Since the integers are infinite, there is a number $Q + 1$.

$Q + 1$ is a number that is not a multiple of any of the primes smaller than $p_z$ and, as such, must be a multiple of a prime greater than $p_z$ (since all integers can be expressed as a product of primes).

This disproves the assumption that there is a greatest prime number, $p_z$.

Therefore there are an infinite number of primes.

If a question asks for proof by contradiction, you should use it. However, if the question doesn't specify the method, think carefully which type of proof will be the simplest and most compelling.

> **Example**
> Prove that if $n^5 + 3$ is even, then $n$ must be odd.
>
> Either a proof by deduction or a proof by contradiction could be used.
>
> **Proof by deduction:**
>
> $n^5 + 3$ is even so $n^5 + 3 = 2C$
> $$n^5 = 2C - 3$$
>
> $\therefore n^5$ is odd
>
> If $n$ is odd, it can be written as $2k + 1$.
> $$(2k + 1)^5 = 32k^5 + 80k^4 + 80k^3 + 40k^2 + 10k + 1$$
> (binomial expansion)
> $$= 2(16k^5 + 40k^4 + 40k^3 + 20k^2 + 5k) + 1,$$
> which is odd.
>
> If $n$ is even, it can be written as $2j$.
> $$(2j)^5 = 32j^5, \text{ which is even.}$$
>
> For $n^5$ to be odd, $n$ must be odd.
>
> **Proof by contradiction:**
>
> Assume that if $n$ is even, then $n^5 + 3$ is even.
>
> Let $n = 2k$
>
> $$n^5 + 3 = (2k)^5 + 3 = 32k^5 + 3 = 2(16k^5 + 1) + 1$$
>
> This expression is odd, so the original assumption is disproved.
>
> Therefore if $n^5 + 3$ is even, $n$ must be odd.

## Don't forget

Sometimes a proof can seem to be so self-evident that 'because it is' feels like an appropriate response. In these cases, it is even more important to concentrate on the small steps and include all assumptions clearly. Without all the steps, the proof is meaningless.

**Links to Other Concepts**
- Algebraic manipulation
- Number theory and sets • Factorisation
- Indices • Problem solving
- Polynomials • Quadratics

## Graphs and Diagrams

You can use graphs and diagrams as part of proofs. Using a calculator with a graphing function can help to simplify things but it can also distract you. Don't expect the calculator to do the thinking for you and, if you are starting to use a trial and improvement approach, it could take some time. Detailed sketches are required to support working.

### QUICK TEST

1. What method of proof would be used to show that $\sqrt{3}$ is an irrational number?

2. What would be the first step in proving that $a \leqslant \sqrt{c}$ or $b \leqslant \sqrt{c}$, where $a$, $b$ and $c \in \mathbb{R}^+$ and $ab = c$.

## SUMMARY

- You must be able to use and interpret set notation. Using appropriate notation helps to make working clear and concise.

- Types of proof required at A-level:

  - **Proof by deduction:** a set of logical steps taken to get to the desired conclusion.

  - **Proof by exhaustion:** all possible values are tried against the rule and, if all work, then the rule is proved.

  - **Disproof by counter-example:** by finding one example that doesn't fit the rule, the rule is disproved.

  - **Proof by contradiction:** requires an assumption that something is true and logical steps to disprove it.

- Showing all steps in a clear way supports the logical approach and helps others to interpret the work done. Use 'let', 'hence' and 'assuming' where relevant. Don't skip steps even if they seem obvious.

## PRACTICE QUESTIONS

1. Billie says that all the students in her class with blue eyes are shorter than those with 'non-blue' eyes.

   All the students are measured and their eye colour is recorded along with their height. The students are labelled A–O in this table:

   | | | |
   |---|---|---|
   | A: Blue 173.5 cm | F: Blue 174.5 cm | K: Brown 180 cm |
   | B: Brown 183 cm | G: Brown 176 cm | L: Brown 180.5 cm |
   | C: Hazel 177.5 cm | H: Blue 168 cm | M: Green 174.5 cm |
   | D: Blue 170 cm | I: Brown 179.5 cm | N: Hazel 178.5 cm |
   | E: Grey 177 cm | J: Greeny brown 176.5 cm | O: Brown 179 cm |

   Show that Billie is wrong. **[3 marks]**

2. Given that $a$ and $b \in \mathbb{Z}$, prove that $a^2 - 4b \neq 2$. **[6 marks]**

# Differentiation

## Note on Notation

$$f'(x) \qquad \frac{dy}{dx} \qquad \frac{d}{dx}f(x)$$

These all mean the **differential** (or **derivative**) of a function.

## Differentiation from First Principles

It is important to be able to differentiate from first principles for small positive powers of $x$.

> **Example**
> Differentiate $y = x^3$ from first principles.
>
> A point $(x, y)$ is taken on the curve. A second point is taken very close to the first, with coordinates $(x + \delta x, y + \delta y)$.
>
> $$y + \delta y = (x + \delta x)^3$$
> $$= x^3 + 3x^2(\delta x) + 3x(\delta x)^2 + (\delta x)^3$$
>
> (Bracket expanded using binomial expansion or by multiplying out brackets)
>
> $$\delta y = 3x^2(\delta x) + 3x(\delta x)^2 + (\delta x)^3$$
>
> The gradient is $\dfrac{\delta y}{\delta x} = \dfrac{3x^2(\delta x) + 3x(\delta x)^2 + (\delta x)^3}{\delta x}$
>
> $$= 3x^2 + 3x(\delta x) + (\delta x)^2$$
>
> Gradient of the tangent $\dfrac{dy}{dx} = \lim\limits_{\delta x \to 0} \dfrac{\delta y}{\delta x}$
>
> $$= \lim\limits_{\delta x \to 0} 3x^2 + 3x(\delta x) + (\delta x)^2 = 3x^2$$

You will need to show differentiation from first principles for both functions $y = \sin x$ and $y = \cos x$.

## Algebraic Differentiation of Polynomials

Differentiation is an algebraic method used to find the gradient function $f'(x)$ of a given function $f(x)$.

If considering a single term from a polynomial, i.e. $ax^b$, the differential of the term is $(ab)x^{b-1}$.

When a polynomial has multiple terms, each term is differentiated independently and the answer is the sum of the differentiated terms. Watch out for negatives and remember that $x = x^1$ and $2 = 2x^0$.

> **Example**
> Differentiate $y = 3x^2 - 5x + 2$.
>
> $$\frac{dy}{dx} = 2 \times 3x^{2-1} - 1 \times 5x^{1-1} + 0 \times 2 = 6x - 5$$

The reason the constant ($+2$) doesn't affect the gradient function can be linked with the graph transformations. The graphs shown are vertical translations of each other, but at each $x$-value the gradient of the graphs is the same.

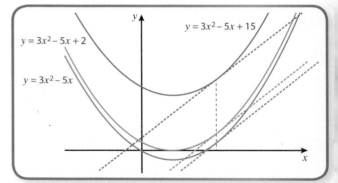

## Non-Integer Indices

A question can appear harder when the index is hidden within alternate notation. Often, the first step is to expand brackets and write each term in simplified index form.

> **Example**
> Find the gradient function of $y = (\sqrt{x} + 2)\left(\frac{1}{x} - 2\right)$.
>
> Expand the brackets or convert to index form:
>
> $$(\sqrt{x} + 2)\left(\frac{1}{x} - 2\right) = \frac{1}{\sqrt{x}} + \frac{2}{x} - 2\sqrt{x} - 4$$
>
> By using the index laws, this can be rewritten to make it easy to differentiate:
>
> $$= x^{-\frac{1}{2}} + 2x^{-1} - 2x^{\frac{1}{2}} - 4$$
>
> Now that each term is in index form, each of them is multiplied by the power, then has 1 taken from the power:
>
> $$\frac{dy}{dx} = -\frac{1}{2} \times x^{\left(-\frac{1}{2}-1\right)} + (-1) \times 2x^{(-1-1)} - \frac{1}{2} \times 2x^{\frac{1}{2}-1} - 0 \times 4$$
>
> $$= -\frac{1}{2}x^{-\frac{3}{2}} - 2x^{-2} - x^{-\frac{1}{2}}$$

# Differentiating More Complex Functions

When differentiating functions, or terms in functions, that aren't of the form $ax^n$, the following should be recognised and learned by heart:

$$\frac{d(e^{kx})}{dx} = ke^{kx} \qquad \frac{d(\ln x)}{dx} = \frac{1}{x}$$

$$\frac{d(\sin kx)}{dx} = k\cos kx \qquad \frac{d(\cos kx)}{dx} = -k\sin kx$$

Some standard results are provided in the formulae booklets. You should be able to recognise and apply these.

$$\frac{d(\tan kx)}{dx} = k\sec^2 kx$$

### Example

Find $\frac{dy}{dx}$, where $y = 2e^{\frac{3x}{2}} + 5\ln x - \cos(-2x)$

$$\frac{dy}{dx} = 2 \times \frac{3}{2}e^{\frac{3}{2}x} + 5 \times \frac{1}{x} - ((-2)(-\sin(-2x)))$$

$$= 3e^{\frac{3}{2}x} + \frac{5}{x} - 2\sin(-2x)$$

As before, each term can be differentiated independently and any coefficient to a term (e.g. the 5 in front of the $\ln x$ above) will multiply the new coefficient of the differentiated term. As ever, be very careful with negatives.

If necessary, you may be able to manipulate these terms into a differentiable form using skills from trigonometry, exponentials and logs. If asked to differentiate something that doesn't fit the form, consider changing the form as a first step.

### Example

Differentiate $f(x) = \ln x^3 + \sin 4x \sec 4x$

$$f(x) = 3\ln x + \sin 4x \frac{1}{\cos 4x} = 3\ln x + \tan 4x$$

$$f'(x) = \frac{3}{x} + 4\sec^2 4x$$

# Finding the Gradient at a Point

Having found the gradient function, the gradient can be found for any given $x$-value by substituting it in. It could also be used to find the coordinate of a point on the graph with any given gradient.

### Example

Given that $f(x) = 6x^{\frac{4}{3}} + 4x - 2$, find the coordinates of the point on the curve, $f(x)$, for which $f'(x) = 10$.

$$f'(x) = 8x^{\frac{1}{3}} + 4$$

$$8x^{\frac{1}{3}} + 4 = 10$$

$$8x^{\frac{1}{3}} = 6 \qquad \therefore x^{\frac{1}{3}} = \frac{3}{4}$$

$$x = \frac{27}{64}$$

To find the $y$-coordinate, substitute back into $f(x)$:

$$6 \times \left(\frac{27}{64}\right)^{\frac{4}{3}} + 4 \times \left(\frac{27}{64}\right) - 2 = \frac{203}{128}$$

$$\left(\frac{27}{64}, \frac{203}{128}\right)$$

# Increasing and Decreasing Functions

To determine if a function is increasing or decreasing at a given point, find $\frac{dy}{dx}$ and substitute in the given $x$-value to find the gradient at that point:

● If $\frac{dy}{dx} > 0$, then the function is **increasing**.

● If $\frac{dy}{dx} < 0$, then the function is **decreasing**.

● If $\frac{dy}{dx} = 0$, then the function is **stationary**.

To change from an increasing function to a decreasing function, the graph must either have a stationary point (true for all continuous graphs) or pass an asymptote. The gradient function will have the same vertical asymptotes as the function since the gradient of an undefined point on a graph is undefined.

### Example
a) Determine if the function $f(x) = 2x^3 + 3x^2 - 12x + 1$ is increasing, decreasing or stationary at points:

(1, −6)     (−3, 10)     (2, 5)     (−1, 14)

The first step is to find $f'(x)$ in each case.

$$f'(x) = 6x^2 + 6x - 12$$
$$f'(1) = (6 \times 1^2) + (6 \times 1) - 12 = 0$$
At (1, −6) the graph is stationary.

$$f'(-3) = 6 \times (-3)^2 + 6 \times (-3) - 12 = 24$$
At (−3, 10) the graph is increasing.

$f'(2) = (6 \times 2^2) + (6 \times 2) - 12 = 24$
At $(2, 5)$ the graph is increasing.

$f'(-1) = 6 \times (-1)^2 + 6 \times (-1) - 12 = -12$
At $(-1, 14)$ the graph is decreasing.

## Finding Stationary Points

To find a stationary point, find the gradient function $\left(\dfrac{dy}{dx}\right)$ and set it equal to zero and solve for $x$. If asked to find the coordinates of the stationary point, substitute the value found for $x$ back into the original equation to find the $y$-value.

### Example
The function $f(x) = 2x^3 + 3x^2 - 12x + 1$ has one stationary point at $(1, -6)$. Find the coordinates of the second stationary point.

$$f'(x) = 6x^2 + 6x - 12$$
$$6x^2 + 6x - 12 = 0$$
$$6(x^2 + x - 2) = 0$$
$$x^2 + x - 2 = 0$$
$$(x-1)(x+2) = 0$$

Having already found the stationary point at $x = 1$, the second point is at $x = -2$.

$$f(-2) = 2(-2)^3 + 3(-2)^2 - 12(-2) + 1 = 21$$

The second stationary point is at $(-2, 21)$.

## The Nature of Stationary Points

Stationary points, or turning points, occur when the gradient of a curve momentarily equals zero.

To determine the nature of the turning point (i.e. whether it is a maximum, a minimum or a point of inflection), the **second differential** $\left(f''(x) \text{ or } \dfrac{d^2y}{dx^2}\right)$ can be taken.

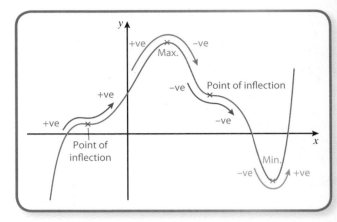

At a **minimum** the gradient is 0 as it changes from a negative to a positive gradient. This is implied by a positive second differential (the overall change from negative to positive is a positive change).
**If $f'(x) = 0$ and $f''(x) > 0$, the stationary point is a minimum.**

At a **maximum** the gradient is 0 as it changes from a positive to a negative gradient. This is implied by a negative second differential (the overall change from positive to negative is a negative change).
**If $f'(x) = 0$ and $f''(x) < 0$, the stationary point is a maximum.**

At a **point of inflection** there is a change in the curvature of the graph. It is a point when the second differential equals 0. Note: the second differential being 0 is not enough to imply that it is a point of inflection. It may represent multiple turning points occurring together.

### Example
The function $f(x) = 2x^3 + 3x^2 - 12x + 1$ has stationary points at $(-2, 21)$ and $(1, -6)$. Determine the nature of each of these turning points.

First find $f''(x)$: $\quad f'(x) = 6x^2 + 6x - 12$
$$f''(x) = 12x + 6$$
At $(-2, 21)$, $f''(x) = 12 \times (-2) + 6 = -18$

$-18 < 0 \therefore$ the stationary point is a maximum.

At $(1, -6)$, $f''(x) = 12 \times (1) + 6 = 18$

$18 > 0 \therefore$ the stationary point is a minimum.

Note: A minimum found this way could be the overall minimum value for the graph but could also be a local minimum (where the curve makes a U shape but there are other values of $x$ that would give a smaller value for $y$).

Points of inflection can have a first differential of 0, making it a stationary point of inflection. If the first differential $f'(x) \neq 0$ but the second differential does, $f''(x) = 0$, then the point of inflection (assuming it is one) marks the change between a concave section and a convex section of the graph.

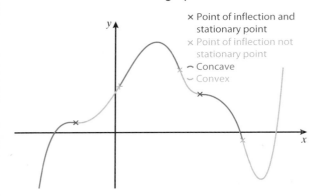

× Point of inflection and stationary point
× Point of inflection not stationary point
⌢ Concave
⌣ Convex

● In a **concave** section of a graph, $f''(x) \leqslant 0$. The range of values for which a graph is concave is $[a, b]$ where $f''(x) \leqslant 0$ for all $x \in [a, b]$.

● In a **convex** section of a graph, $f''(x) \geqslant 0$. The range of values for which a graph is convex is $[a, b]$ where $f''(x) \geqslant 0$.

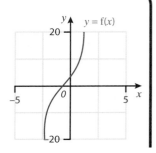
## Tangents, Normals and Coordinates

By finding gradients of curves, you can find the gradient of the tangent to the curve at any given point, and hence find the gradient of the normal to the curve.

Finding the equation of a straight line: $y_2 - y_1 = m(x_2 - x_1)$.

To use this equation, you need:

1. A point $(x_1, y_1)$ that the line passes through.
2. Either a second point (which can be used to find the gradient) or the gradient of the line ($m$).

At $x = -1$, $\frac{dy}{dx} = 9 - 12 = -3$

The gradient of the tangent $m = -3$.

$y - y_1 = m(x - x_1)$
$y - 9 = -3(x - -1)$
$y - 9 = -3x - 3$
$\quad\ y = -3x + 6$

**b)** Find the coordinates of the point where the tangent intercepts the curve.

To find intersections, set the equations equal to each other:

$-3x + 6 = 3x^3 - 12x$
$3x^3 - 9x - 6 = 0$
$\ x^3 - 3x - 2 = 0$

This gives a cubic so there will be three roots to the equation. Sketching the original graphs will help:

$y = 3x^3 - 12x =$
$x(3x^2 - 12) =$
$x(3x - 6)(x + 2)$,
so positive cubic
curve with roots
at $x = -2$, $x = 0$
and $x = 2$.

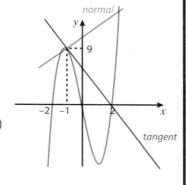

Tangent at $(-1, 9)$
has negative
gradient so is to
the right of the
local maximum.

Line $y = -3x + 6$ has $x$-intercept at 2.

The answer to the equation $x^3 - 3x - 2 = 0$ will have a repeat root at $(-1, 9)$, as it touches but does not cross the line here. The other root is at $(2, 0)$, by inspection of graphs.

Check these are **all** the roots to the equation $x^3 - 3x - 2 = 0$ by expansion (alternate methods could be used):

$(x + 1)^2(x - 2) = (x^2 + 2x + 1)(x - 2)$
$\qquad\qquad = x^3 + 2x^2 - 2x^2 + x - 4x - 2$
$\qquad\qquad = x^3 - 3x - 2$

The point where the tangent crosses the curve is at $x = 2$.

$y = (3 \times 2^3) - (12 \times 2) = 0$

The point is $(2, 0)$.

**c)** Find the exact greatest distance between any two points where the normal intersects the curve.

The gradient of the normal is $-\frac{1}{-3} = \frac{1}{3}$ and it passes through the point $(-1, 9)$.

Equation of the normal $y - y_1 = m(x - x_1)$
$y - 9 = \frac{1}{3}(x - -1)$
$3y - 27 = x + 1$
$3y = x + 28$
$y = \frac{1}{3}x + \frac{28}{3}$

Find the points of intersection with the curve by setting them equal to each other:

$3x^3 - 12x = \frac{1}{3}x + \frac{28}{3}$
$3(3x^3 - 12x) = x + 28$
$9x^3 - 36x = x + 28$
$9x^3 - 37x - 28 = 0$

In this case you know one solution already, $x = -1$, which means that $(x + 1)$ is a factor.

$$\begin{array}{r}
9x^2 - 9x - 28 \\
x+1 \overline{\smash{\big)}\ 9x^3 + 0x^2 - 37x - 28} \\
\underline{9x^3 + 9x^2} \phantom{xxxxxxxxx} \\
-9x^2 - 37x \phantom{xx} \\
\underline{-9x^2 - 9x} \phantom{xx} \\
-28x - 28 \\
\underline{-28x - 28}
\end{array}$$

The other two intersections will be found by solving $9x^2 - 9x - 28 = 0$:

$x = \frac{-b \pm \sqrt{b^2 - 4ac}}{2a}$

$x = \frac{9 \pm \sqrt{(-9)^2 - 4 \times 9 \times -28}}{2 \times 9}$

$x = \frac{7}{3} \quad \rightarrow \quad y = \frac{1}{3} \times \frac{7}{3} + \frac{28}{3} = \frac{91}{9}$

$x = -\frac{4}{3} \quad \rightarrow \quad y = \frac{1}{3} \times \frac{-4}{3} + \frac{28}{3} = \frac{80}{9}$

Find distance using Pythagoras' Theorem:

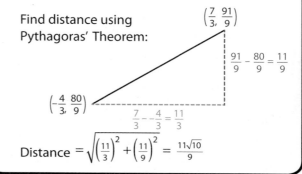

Distance $= \sqrt{\left(\frac{11}{3}\right)^2 + \left(\frac{11}{9}\right)^2} = \frac{11\sqrt{10}}{9}$

## SUMMARY

- $f'(x) = \lim\limits_{\delta x \to 0} \dfrac{f(x + \delta x) - f(x)}{\delta x}$

- You need to know the differentials of $x^a$, $e^{kx}$, $\ln x$, $\sin kx$, $\cos kx$, $\tan kx$ and $a^{kx}$.

- Use index notation to get an expression into a differentiable form.

- Expand brackets and simplify before differentiating.

- To find a gradient at a given point, find the gradient function $\left(\dfrac{dy}{dx}\right)$ and substitute the value of $x$ for which you want the gradient.

- If $\dfrac{dy}{dx} > 0$, the function is increasing.

- If $\dfrac{dy}{dx} < 0$, the function is decreasing.

- If $\dfrac{dy}{dx} = 0$, the function is stationary.

- To find the coordinates of a stationary point:
  - differentiate, set equal to zero and solve
  - find the $y$-coordinate by substituting the $x$-value(s) back into the original equation.

- $f''(x)$ or $\dfrac{d^2y}{dx^2}$ means the second differential. It is the rate of change of the gradient.

- If $f'(x) = 0$ and $f''(x) > 0$, the stationary point is a minimum. If $f'(x) = 0$ and $f''(x) < 0$, the stationary point is a maximum.

- A point of inflection has $f''(x) = 0$. But $f''(x) = 0$ doesn't imply a point of inflection:
  - Check the behaviour of the curve either side; if there is a change in curvature, it is a point of inflection.
  - In a concave section of a graph, $f''(x) \leq 0$.
  - In a convex section of a graph, $f''(x) \geq 0$.

- Tangents have the same gradient as the function at a given point.

- Normals have the negative reciprocal of the gradient of the tangent.

- To find a point where a curve is intersected by a normal or a tangent, one result is already given and so can be divided out.

## Links to Other Concepts

- Polynomials
- Coordinate geometry
- Algebraic manipulation
- Indices
- Mechanics
- Integration
- Binomial expansion
- Maximisation/optimisation problems
- Growth and decay
- Logarithms
- and exponentials
- Quadratics, factor theorem

## QUICK TEST

1. A curve has an equation $y = f(x)$, where $f(x) = 3\sin x + e^x + 2$, shown on the graph in green. The red curve is the plot of $y = f''(x)$. All the points of inflection are in the range $x < 0$.

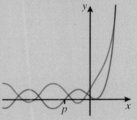

   a) Find the equation for the red curve.

   b) How many points of inflection are shown on the graph for the curve $y = f(x)$?

   c) For $f(x)$, is the region $x \geq p$ convex or concave? Give a reason for your answer.

## PRACTICE QUESTIONS

1. $y = (\sqrt[3]{x^2} + 4e^{2x})((\sqrt[3]{x})^2 - (2e^x)^2)$

   a) Expand and simplify the expression, leaving all terms in index form. **[3 marks]**

   b) Find $\dfrac{dy}{dx}$ **[2 marks]**

   c) Hence find the gradient in terms of e, when $x = 8$. **[2 marks]**

2. a) Find the gradient function of the curve $y = \dfrac{\ln x}{10} - 12x^5 + e^{3x}$. **[3 marks]**

   b) A line running parallel to the tangent at $x = 1$ has the equation $y = ax + 2$. Find the exact value of $a$. **[2 marks]**

   c) Show that there is a local maximum between $x = 1$ and $x = 1.1$. **[4 marks]**

63

# Differentiation and Combined Functions

## Chain Rule

The chain rule is used when the expression to be differentiated is a function within a function.

When $y = f(g(x))$, let $u = g(x)$ and so $y = f(u)$

$$\frac{dy}{dx} = \frac{dy}{du} \times \frac{du}{dx}$$

### Example

Find $\dfrac{dy}{dx}$ when $y = \dfrac{(3x^2 + 2x)^{\frac{4}{3}}}{5}$

Let $u = 3x^2 + 2x \quad \rightarrow \quad \dfrac{du}{dx} = 6x + 2$

$y = \dfrac{1}{5}u^{\frac{4}{3}} \quad \rightarrow \quad \dfrac{dy}{du} = \dfrac{4}{15}u^{\frac{1}{3}}$

$\dfrac{dy}{dx} = \dfrac{4}{15}u^{\frac{1}{3}}(6x + 2) = \dfrac{4}{15}(3x^2 + 2x)^{\frac{1}{3}}(6x + 2)$

## Product Rule

The product rule is used when the expression to be differentiated is the product of two functions.

$$\frac{d}{dx}(f(x)g(x)) = f'(x)g(x) + f(x)g'(x)$$

### Example

$y = 2x^3 \sin x$. Find the exact gradient of the curve when $x = \pi$.

Let $f(x) = 2x^3 \quad \rightarrow \quad f'(x) = 6x^2$

Let $g(x) = \sin x \quad \rightarrow \quad g'(x) = \cos x$

$\dfrac{dy}{dx} = 6x^2 \sin x + 2x^3 \cos x$

At $x = \pi$, $\dfrac{dy}{dx} = 6\pi^2 \sin \pi + 2\pi^3 \cos \pi$

$\qquad = 6\pi^2 \times 0 + 2\pi^3 \times -1$

$\qquad = -2\pi^3$

## Quotient Rule

The quotient rule is given in formulae booklets. It is used when the expression to be differentiated is one function divided by another function.

$$\frac{d}{dx}\left(\frac{f(x)}{g(x)}\right) = \frac{f'(x)g(x) - f(x)g'(x)}{(g(x))^2}$$

### Example

Use $\tan x \equiv \dfrac{\sin x}{\cos x}$ to prove that $\dfrac{d}{dx}(\tan x) = 1 + \tan^2 x$.

Let $f(x) = \sin x \quad \rightarrow \quad f'(x) = \cos x$

Let $g(x) = \cos x \quad \rightarrow \quad g'(x) = -\sin x$

$\dfrac{d}{dx}\left(\dfrac{\sin x}{\cos x}\right) = \dfrac{\cos x \cos x - \sin x(-\sin x)}{\cos^2 x}$

$\qquad = \dfrac{\cos^2 x}{\cos^2 x} + \dfrac{\sin^2 x}{\cos^2 x}$

$\qquad = 1 + \tan^2 x$

You can use the quotient rule and the product rule interchangeably, with a little rearrangement/algebraic manipulation or in combination with the chain rule.

If asked to find $\dfrac{d}{dx}\left(5x^{\frac{2}{5}} \times \ln x\right)$, the product rule could be applied directly but, if needed, it could be considered as $\dfrac{d}{dx}\left(\dfrac{5\ln x}{x^{-\frac{2}{5}}}\right)$ in order to apply the quotient rule.

## Inverses

The inverse of $\dfrac{dy}{dx}$ is $\dfrac{1}{\frac{dy}{dx}} = \dfrac{dx}{dy}$ and the inverse

of $\dfrac{dx}{dy}$ is $\dfrac{1}{\frac{dx}{dy}} = \dfrac{dy}{dx}$

To find $\left(\dfrac{dx}{dy}\right)$, either invert the solution to $\left(\dfrac{dy}{dx}\right)$, or define $x$ in terms of $y$ and follow the rules of differentiation but with $x$ and $y$ 'swapped around'.

This is useful where you are asked to find the gradient of a curve $\left(\dfrac{dy}{dx}\right)$ but are given an expression where it is impossible to rearrange to give $y$ in terms of $x$.

---

**Example**

Find the equation of the tangent to the line $x + \ln y^3 = y^2 - 4$, at the point where $y = 2$, in the form $y = ax + \ln 2^b + c$.

$x = y^2 - 4 - 3\ln y$      Passes through $(-3\ln 2, 2)$

$\dfrac{dx}{dy} = 2y - \dfrac{3}{y}$

$\dfrac{dy}{dx} = \dfrac{1}{2y - \dfrac{3}{y}}$     Gradient $\left(\dfrac{dy}{dx}\right)$ at $y = 2$ is $\dfrac{2}{5}$

Equation of tangent: $y - y_1 = m(x - x_1)$

$$y - 2 = \tfrac{2}{5}(x - (-3\ln 2))$$

$$y = \tfrac{2}{5}x + \tfrac{6}{5}\ln 2 + 2$$

$$y = \tfrac{2}{5}x + \ln 2^{\frac{6}{5}} + 2$$

---

# Differentiating Parametric Equations

If a relationship is defined in terms of parametric equations, you can differentiate and then combine them. To do this, the chain rule is used:

$$\frac{dy}{dx} = \frac{dy}{dt} \times \frac{dt}{dx} = \frac{\left(\dfrac{dy}{dt}\right)}{\left(\dfrac{dx}{dt}\right)}$$

Or given that $y = f(t)$ and $x = g(t)$

$$\frac{dy}{dx} = f'(t) \times \frac{1}{g'(t)}$$

---

**Example**

The graph shows the curve $x = \sin t$, $y = t\cos 2t$, where $t \in [-15, 15]$. The red line is the tangent to point P, where $t = \dfrac{\pi}{6}$. Find the value of $q$.

---

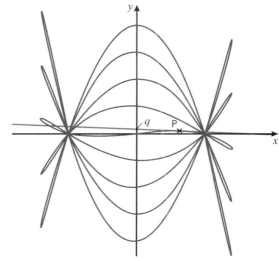

$\dfrac{dx}{dt} = \cos t \; \therefore \; \dfrac{dt}{dx} = \dfrac{1}{\cos t}$

To find $\dfrac{dy}{dt}$ the product rule is needed:

$$\frac{d}{dx}(f(x)g(x)) = f'(x)g(x) + f(x)g'(x)$$

Let $f(x) = t$      $\rightarrow$    $f'(x) = 1$

Let $g(x) = \cos 2t$    $\rightarrow$    $g'(x) = -2\sin 2t$

$$\frac{dy}{dt} = \cos 2t - 2t\sin(2t)$$

$$\frac{dy}{dx} = (\cos 2t - 2t\sin(2t)) \times \left(\frac{1}{\cos t}\right)$$

At $t = \dfrac{\pi}{6}$    $\dfrac{dy}{dx} = \dfrac{\cos\dfrac{\pi}{3} - \dfrac{\pi}{3}\sin\left(\dfrac{\pi}{3}\right)}{\cos\dfrac{\pi}{6}}$

$$= \frac{\dfrac{1}{2} - \dfrac{\pi}{3} \times \dfrac{\sqrt{3}}{2}}{\dfrac{\sqrt{3}}{2}} = \frac{1}{\sqrt{3}} - \frac{\pi}{3}$$

Passes through the point $P\left(\dfrac{1}{2}, \dfrac{\pi}{12}\right)$

$$\left(x = \sin\frac{\pi}{6} \text{ and } y = \frac{\pi}{6} \times \cos\frac{2\pi}{6}\right)$$

$$y - \frac{\pi}{12} = \left(\frac{1}{\sqrt{3}} - \frac{\pi}{3}\right)\left(x - \frac{1}{2}\right)$$

You can multiply this out fully but, since the question asks for the $y$-intercept, the constant is what matters:

$$q = y\text{-intercept} = -\frac{1}{2}\left(\frac{1}{\sqrt{3}} - \frac{\pi}{3}\right) + \frac{\pi}{12} = \frac{\pi}{4} - \frac{\sqrt{3}}{6}$$

## Implicit Differentiation

Implicit differentiation can be used if a relationship between $x$ and $y$ is defined but unable to be given easily with one variable in terms of the other. The product rule and the chain rule are used, and inverses may be relevant. Take each term at a time and differentiate it. Then look to rearrange the resulting equation to give an expression (in terms of $x$ and $y$) for $\frac{dy}{dx}$.

**Example**

$y^2 + x^2 + 6y + xy - 12 = 0$. Find $\frac{dy}{dx}$ and hence the gradient(s) of the graph at the points where $x = 2$.

Each term can be considered separately:

$\frac{d}{dx}(y^2) = \frac{dy}{dx} \times 2y$ (by chain rule, let $y = u$, giving

$\qquad \frac{du}{dx} = \frac{dy}{dx}$ and $\frac{d(u^2)}{du} = 2u$)

$\frac{d}{dx}(x^2) = 2x$ (by basic differentiation of $ax^n$)

$\frac{d}{dx}(6y) = \frac{dy}{dx} \times 6$ (by chain rule, let $y = u$, giving

$\qquad \frac{du}{dx} = \frac{dy}{dx}$ and $\frac{d(6u)}{du} = 6$)

$\frac{d}{dx}(xy) = x\frac{dy}{dx} + y$ (by product rule where

$\qquad$ $f(x) = x$, $f'(x) = 1$ and

$\qquad$ $g(x) = y$, $g'(x) = \frac{dy}{dx}$)

$\frac{d}{dx}(-12) = 0$ (by basic differentiation of $ax^0$)

$\frac{d}{dx}(0) = 0$ (by basic differentiation of $ax^0$)

Rearranging the implicitly differentiated equation:

$2y\frac{dy}{dx} + 2x + 6\frac{dy}{dx} + x\frac{dy}{dx} + y = 0$

$2y\frac{dy}{dx} + 6\frac{dy}{dx} + x\frac{dy}{dx} = -2x - y$

$\frac{dy}{dx}(2y + 6 + x) = -2x - y$

$\frac{dy}{dx} = \frac{-2x - y}{(2y + 6 + x)}$

At $x = 2$ $\qquad y^2 + 4 + 6y + 2y - 12 = 0$

$\qquad\qquad y^2 + 8y - 8 = 0$

$\qquad\qquad y = \frac{-8 \pm \sqrt{8^2 - 4(1)(-8)}}{2}$

$y = -4 + 2\sqrt{6}$ and $y = -4 - 2\sqrt{6}$

At $(2, -4 + 2\sqrt{6})$, $\frac{dy}{dx} = \frac{-4 - (-4 + 2\sqrt{6})}{2(-4 + 2\sqrt{6}) + 6 + 2} = \frac{-2\sqrt{6}}{4\sqrt{6}} = -\frac{1}{2}$

At $(2, -4 - 2\sqrt{6})$, $\frac{dy}{dx} = \frac{-4 - (-4 - 2\sqrt{6})}{2(-4 - 2\sqrt{6}) + 6 + 2} = \frac{+2\sqrt{6}}{-4\sqrt{6}} = -\frac{1}{2}$

## Interpreting Questions in Context

Questions will often be given in the context of a rate of change. Set up the equations from the information and define the terms you use if none are given.

**Example**

An open-top box is to be made from a rectangle of card that measures three times as long as its width, $a$. By cutting a square from each corner, the sides are then folded up to make the open box. Each square is $x$ cm wide.

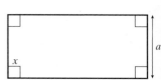

**a)** Show that the volume of the box is $V = 3a^2x - 8ax^2 + 4x^3$.

$V = b \times w \times h$

$V = (3a - 2x)(a - 2x)(x)$

$V = (3a^2 - 2ax - 6ax + 4x^2)x$

$V = 3a^2x - 8ax^2 + 4x^3$

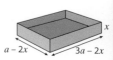

**b)** In terms of $a$, find the value for $x$ for which the box will have the largest volume.

$\frac{dV}{dx} = 3a^2 - 16ax + 12x^2$

Turning point when $\frac{dV}{dx} = 0$:

$0 = 3a^2 - 16ax + 12x^2$

$x = \frac{-b \pm \sqrt{b^2 - 4ac}}{2a}$

$x = \frac{16a \pm \sqrt{(-16a)^2 - 4 \times 12 \times 3a^2}}{2 \times 12}$

$= \frac{16a \pm \sqrt{256a^2 - 144a^2}}{24} = \frac{16a \pm \sqrt{112a^2}}{24}$

$x = \frac{(16 + 4\sqrt{7})a}{24}$ or $x = \frac{(16 - 4\sqrt{7})a}{24}$

Identifying the maximum point using the second differential: $\dfrac{d^2V}{dx^2} = 24x - 16a$

For the point to be a maximum, $\dfrac{d^2V}{dx^2} < 0$:

$$24x - 16a < 0 \therefore 24x < 16a$$
$$x < \frac{16a}{24}$$

The maximum point is at $x = \dfrac{(16 - 4\sqrt{7})a}{24}$

## SUMMARY

- **Chain rule:** $\dfrac{dy}{dx} = \dfrac{dy}{du} \times \dfrac{du}{dx}$

- **Product rule:** $\dfrac{d}{dx}(f(x)g(x)) = f'(x)g(x) + f(x)g'(x)$

- **The quotient rule is given in the formulae booklets.**

- **Inverses can be used to find gradients where $x$ is defined in terms of a function of $y$.**

- **To differentiate parametric equations, use**

$$\frac{dy}{dx} = \frac{dy}{dt} \times \frac{dt}{dx} = \frac{\left(\frac{dy}{dt}\right)}{\left(\frac{dx}{dt}\right)}$$

- **To differentiate implicitly, use the chain rule and the product rule to differentiate every term of an equation. Rearrange as necessary to find an expression for $\dfrac{dy}{dx}$ or substitute in values given to solve the problem.**

### Links to Other Concepts
- Coordinate geometry  ● Polynomials
- Algebraic manipulation
- Indices  ● Mechanics, rates of change
- Integration  ● Growth and decay
- Quadratics, factor theorem
- Trigonometry and identities
- Binomial expansion

## QUICK TEST

1. Differentiate the following equations:

   a)  $y = (2x + 3)^6$ 　　 b)  $y = \dfrac{3}{x}\sin x$

   c)  $y = 6x^2 e^{4x}$ 　　 d)  $y = 3^{2x} + 4x\cos x$

2. Find an expression for $\dfrac{dy}{dx}$ in terms of $x$, $y$ or $x$ and $y$ for these equations and find the value of the gradient when $y = 2$.

   a)  $3x + 6y = y^3 - 3e^y$ 　 b)  $x = \tan y + 3y - 2$

3. Find the expression for the gradient, the full coordinate(s) for the value given and the exact gradient(s) of each curve at the defined point.

   a)  $x^2 + 4y^2 - 26 = 0$, at $x = 1$

   b)  $x = e^{2t-1}$, $y = 3t$, at $t = 0.5$

   c)  $x = 2t - 4$, $y = t^2 + \sin(2t)$, at $x = \dfrac{2\pi - 12}{3}$

   d)  $e^y = 3x^2$, at $x = 1$

   e)  $2xy - 12\sin x = 0$, at $x = \dfrac{\pi}{2}$

4. a)  Find an expression for $\dfrac{dy}{dx}$ for the curve $x = 3e^{2t}$, $y = 4t^2 - 3$.

   b)  Hence find the coordinates of the turning point.

## PRACTICE QUESTIONS

1. If $f(x) = \sin(e^{2x})$, find $f'(x)$. 　　 **[4 marks]**

2. a)  If $y = 2x^2\sin(e^{2x})$, find $\dfrac{dy}{dx}$. 　 **[4 marks]**

   b)  What is the exact gradient at the point $x = \ln\sqrt{\pi}$? Determine if the function is increasing or decreasing at this point, giving a reason for your answer. 　 **[5 marks]**

3. $xy(x - y) = 12$

   a)  Show that $\dfrac{dy}{dx} = \dfrac{y^2 - 2xy}{x^2 - 2xy}$ 　 **[5 marks]**

   b)  Find the $x$-coordinate of the point where the normals to the curve at $x = 4$ intersect with each other. 　 **[11 marks]**

# Integration 1

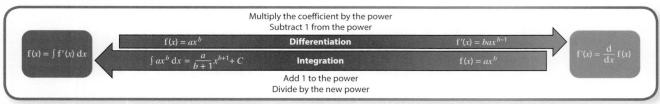

Multiply the coefficient by the power
Subtract 1 from the power

| | |
|---|---|
| $f(x) = ax^b$ | |
| **Differentiation** | $f'(x) = bax^{b-1}$ |

$f(x) = \int f'(x)\,dx$

$\int ax^b\,dx = \frac{a}{b+1}x^{b+1} + C$

**Integration**    $f(x) = ax^b$

$f'(x) = \frac{d}{dx}f(x)$

Add 1 to the power
Divide by the new power

## Integration

Integration is the inverse of differentiation.

A constant $(+ C)$ has to be added during integration, as for every gradient function there is a family of graphs that could represent the original function, all of them vertical translations of each other. The original function $f(x)$ may therefore have contained a constant which differentiated to zero and so 'disappeared' in the gradient function, so a '$+ C$' needs to be added whenever you integrate.

$$\int x^n\,dx = \frac{x^{n+1}}{n+1} + C, \quad n \neq -1$$

### Example

Positive integer power:

$$\int 3x^2\,dx = \frac{3}{2+1}x^{2+1} + C = x^3 + C$$

Power of 1: $\int -4x\,dx = \frac{-4}{1+1}x^{1+1} + C = -2x^2 + C$

Power of 0 (or constant terms):

$$\int 2\,dx = \frac{2}{0+1}x^{0+1} + C = 2x + C$$

Negative integer power:

$$\int 6x^{-2}\,dx = \frac{6}{-2+1}x^{-2+1} + C = -6x^{-1} + C$$

Fractional power:

$$\int 3x^{\frac{5}{2}}\,dx = \frac{3}{\frac{5}{2}+1}x^{\frac{5}{2}+1} + C = \frac{3}{\frac{7}{2}}x^{\frac{7}{2}} + C = \frac{6}{7}x^{\frac{7}{2}} + C$$

As with differentiation, the terms of an expression can be integrated one at a time. Only one $+ C$ is needed, not one for each term.

$$\int ax^n + bx^m\,dx = a\int x^n\,dx + b\int x^m\,dx + C,$$

$$n \neq -1, \; m \neq -1$$

### Example

Integrate the function $y = 12x^4 + 3x^{-\frac{3}{4}} - 3$.

$$\int y\,dx = \int 12x^4 + 3x^{-\frac{3}{4}} - 3\,dx$$

$$= \frac{12}{4+1}x^{4+1} + \frac{3}{-\frac{3}{4}+1}x^{-\frac{3}{4}+1} - \frac{3}{(0+1)}x^{0+1} + C$$

$$= \frac{12}{5}x^5 + 12x^{\frac{1}{4}} - 3x + C$$

## Integrating More Complex Functions

As integration is the inverse of differentiation, these functions can all be integrated:

- $\int e^{kx}\,dx = \frac{1}{k}e^{kx} + C$
- $\int \frac{1}{x}\,dx = \ln x + C$
- $\int \sin kx\,dx = -\frac{1}{k}\cos kx + C$
- $\int \cos kx\,dx = \frac{1}{k}\sin kx + C$

This list is not exhaustive. As with differentiation, some rearrangement or use of identities may be required to gain a form that is easy to integrate.

### Example

Find $\int 14e^{2x} + 5x^{-1} - \sin\frac{x}{10}\,dx$.

$$\int 14e^{2x} + 5x^{-1} - \sin\frac{x}{10}\,dx = \frac{14}{2}e^{2x} + 5\ln x + 10\cos\frac{x}{10} + C$$

## Finding $C$

As all constants are reduced to zero by differentiation, it isn't possible to know the value of $C$ just by integration. To find $C$, extra information is needed. This is generally in the form of a point on the graph $y = f(x)$.

**Example**

$f'(x) = 3x^2 - 4x + 2$

Find $f(x)$ given that when $x = 3$, $f(x) = 7$.

$f(x) = \int 3x^2 - 4x + 2 \, dx$

$\qquad = \frac{3}{3}x^3 - \frac{4}{2}x^2 + 2x + C = x^3 - 2x^2 + 2x + C$

$7 = 3^3 - (2 \times 3^2) + (2 \times 3) + C$

$7 = 27 - 18 + 6 + C$

$7 = 15 + C$

$C = -8$

So $f(x) = x^3 - 2x^2 + 2x - 8$

---

**Example**

The equation $y = e^{4x} + 2x$ is integrated. The graph of the integrated function passes through the origin. Find the expression for $\int y \, dx$.

$$\int y \, dx = \frac{1}{4}e^{4x} + x^2 + C$$

At $x = 0$  $\quad \int y \, dx = \frac{1}{4}e^0 + 0 + C$

$\qquad\qquad\qquad 0 = \frac{1}{4} + C$

$\qquad\qquad\qquad C = -\frac{1}{4}$

$\qquad\qquad \int y \, dx = \frac{1}{4}e^{4x} + x^2 - \frac{1}{4}$

---

## Converting Expressions

As with differentiation, one of the biggest challenges can be getting the expression into a form which can be integrated to start with. Consider expanding brackets using index form.

---

**Example**

$\frac{dy}{dx} = \left(1 + \frac{1}{x\sqrt{x}}\right)^3$. Find an expression for $y$ in terms of $x$.

$\left(1 + \frac{1}{x\sqrt{x}}\right)^3 = \left(1 + x^{-\frac{3}{2}}\right)^3$

$\qquad = 1 + 3 \times x^{-\frac{3}{2}} + 3 \times \left(x^{-\frac{3}{2}}\right)^2 + \left(x^{-\frac{3}{2}}\right)^3$

$\qquad = 1 + 3x^{-\frac{3}{2}} + 3x^{-3} + x^{-\frac{9}{2}}$

$y = \int \left(1 + \frac{1}{\sqrt{x}}\right)^3 dx$

$\qquad = \int 1 \, dx + 3\int x^{-\frac{3}{2}} \, dx + 3\int x^{-3} \, dx + \int x^{-\frac{9}{2}} \, dx$

$y = x - 6x^{-\frac{1}{2}} - \frac{3}{2}x^{-2} - \frac{2}{7}x^{-\frac{7}{2}} + C$

---

**Example**

$y = 7 - 3\sin^2(2x)$

a) Show that $y$ can be written in the form $y = a + b\cos(4x)$, where $a$ and $b$ are constants to be found.

$\qquad y = 4 + (3 - 3\sin^2(2x))$

$\qquad\quad = 4 + 3(1 - \sin^2 2x)$

$\qquad\quad = 4 + 3\cos^2(2x)$

As $\cos 2\theta = 2\cos^2\theta - 1 \rightarrow \cos^2(2x)$

$= \frac{1}{2}(\cos(4x) + 1)$

$\qquad y = 4 + 3\left(\frac{1}{2}(\cos(4x) + 1)\right)$

$\qquad\quad = 4 + \frac{3}{2}\cos(4x) + \frac{3}{2} = \frac{11}{2} + \frac{3}{2}\cos(4x)$

b) Hence find an expression for $\int y \, dx$.

$$\int y \, dx = \frac{11}{2}x + \frac{3}{8}\sin(4x)$$

---

## Using Algebraic Constants

Although much of an integration calculation can be done using a graphical calculator, one way to check your understanding is to use algebraic constants.

If the curve passes through a point $(0, a)$, the constant of integration $C = a$. If a curve passes through the origin, there is no constant term in its equation.

---

**Example**

$f'(x) = 3x^a + 2x^2, \; a \neq -1$

a) Find an expression for $f(x)$.

$\qquad f(x) = \frac{3}{(a + 1)}x^{a+1} + \frac{2}{3}x^3 + C$

b) The curve passes through the points $(1, 2)$ and $\left(0, -\frac{1}{6}\right)$. Find the values of $a$ and $C$.

$\qquad f(0) = -\frac{1}{6} = \frac{3}{(a + 1)} \times 0^{a+1} + \frac{2}{3} \times 0^3 + C = C$

$\qquad C = -\frac{1}{6}$

$\qquad f(1) = 2 = \frac{3}{(a + 1)} \times 1^{a+1} + \frac{2}{3} \times 1^3 - \frac{1}{6}$

$\qquad \frac{3}{(a + 1)} + \frac{2}{3} - \frac{1}{6} = 2$

$\qquad\qquad \frac{3}{a + 1} = 2 + \frac{1}{6} - \frac{2}{3} = \frac{3}{2}$

$\qquad\qquad a + 1 = 2$

$\qquad\qquad\quad a = 1$

## The Meaning of Integration

Differentiating is finding the rate of change, which on a graph is represented by the gradient of the curve. Integrating finds the area between the function and the $x$-axis.

## Between Bounds

The area to be found could be specified as between bounds. A small number is used at the top and bottom of the integration symbol to define the bounds.

$A = \int_a^b f(x)\,dx$ means the integral between $a$ and $b$ of the function $f(x)$ with respect to $x$.

### Example
In this graph, the area is between the $x$-axis, the curve and the lines $x = 1$ and $x = 4$.

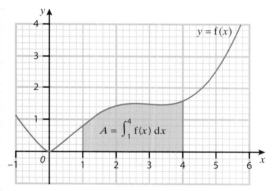

In this graph $A = \int_1^4 f(x)\,dx$.

When finding an integral between bounds, it is important to show the following steps.

### Example
$f(x) = x(x - 4)$. Find the area between the curve, the $x$-axis and the lines $x = 2$ and $x = 1$.

- If necessary, rearrange $f(x)$ into a form that is easily integrated (simplified index form):
  $f(x) = x(x - 4) = x^2 - 4x$

- Sketching the curve may help. Sketches are good for visualisation and for helping to spot unreasonable answers.

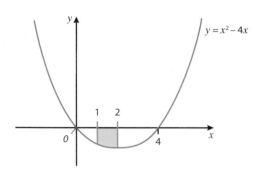

- State the integration to be performed with limits:

  $$A = \int_1^2 x^2 - 4x \; dx$$

- Use square brackets to show the integrated function:

  $$A = \left[ \frac{1}{3}x^3 - 2x^2 \right]_1^2$$

- Show the method for finding the area between bounds, which is $(g(b)) - (g(a))$:

  $$A = \left( \frac{1}{3}(2)^3 - 2(2)^2 \right) - \left( \frac{1}{3}(1)^3 - 2(1)^2 \right)$$

- Solve:

  $$A = \left( -\frac{16}{3} \right) - \left( -\frac{5}{3} \right) = -\frac{11}{3}$$

- Check the answer is reasonable and convert it to positive if it is negative (because it is an area). Keep in mind any context and include units if relevant.

  $$A = \frac{11}{3} \; \text{units}^2$$

### Don't forget

The integrated function **doesn't need to include '+$C$' when between bounds**, as it would cancel out anyway when the two values are subtracted from each other.

## Between Roots

If a question requires the roots to be found, the same process is followed but with a first step of identifying the bounds by solving the equation. Note: If the area is bounded by the curve, the positive $x$-axis and the positive $y$-axis, one of the bounds is $x = 0$ and the other is a root to the equation to be found.

## Example

Find the area enclosed by the curve $y = 3x^2 - 5x - 2$ and the $x$-axis.

The first step is to set equal to zero and solve to find roots. Sketch the curve to show the roots and the shape to help identify the area to be found:

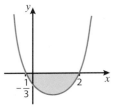

$3x^2 - 5x - 2 = 0$

$(3x + 1)(x - 2) = 0$

$x = -\frac{1}{3}$ and $x = 2$

Now following the steps shown previously:

$A = \int_{-\frac{1}{3}}^{2} (3x^2 - 5x - 2)\ dx$

$= \left[ x^3 - \frac{5}{2}x^2 - 2x \right]_{-\frac{1}{3}}^{2}$

$= \left(2^3 - \frac{5}{2} \times 2^2 - 2 \times 2\right) - \left(\left(-\frac{1}{3}\right)^3 - \frac{5}{2} \times \left(-\frac{1}{3}\right)^2 \right.$

$\left. - 2 \times \left(-\frac{1}{3}\right)\right)$

$= -6 - \frac{19}{54} = -\frac{343}{54}$

Area $= \frac{343}{54}$ units$^2$

## Composite Areas

The individual areas are added or subtracted to work out the required area of a composite shape:

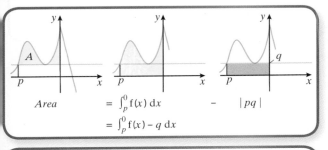

$$Area = \int_{p}^{0} f(x)\ dx \quad - \quad |pq|$$
$$= \int_{p}^{0} f(x) - q\ dx$$

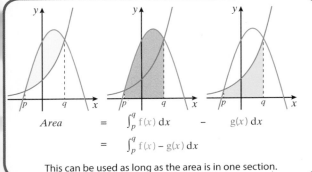

$$Area = \int_{p}^{q} f(x)\ dx \quad - \quad g(x)\ dx$$
$$= \int_{p}^{q} f(x) - g(x)\ dx$$

This can be used as long as the area is in one section.

## Example

The graph shows the curve $y = 3x^5 + x^3 + 2$ and the straight line $y = 4x + 2$.

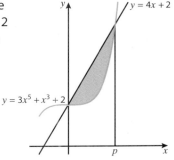

**a)** Find the value of $p$ and the coordinates of the intersection at $x = p$.

$$3x^5 + x^3 + 2 = 4x + 2$$
$$3x^5 + x^3 - 4x = 0$$
$$x(3x^2 + 4)(x^2 - 1) = 0$$
$$x = 0$$
$$3x^2 + 4 = 0 \rightarrow \text{No real results}$$
$$x^2 - 1 = 0 \rightarrow x = 1, x = -1$$
$$x = 0, x = -1, x = 1$$

As $p$ is positive, as shown on the graph, $p = 1$.

At $x = 1 \qquad y = (4 \times 1) + 2 = 6$

Point of intersection is $(1, 6)$.

**b)** Find $\int_{0}^{p} 3x^5 + x^3 + 2\, dx$.

$\int_{0}^{1} 3x^5 + x^3 + 2\ dx = \left[ \frac{1}{2}x^6 + \frac{1}{4}x^4 + 2x \right]_{0}^{1}$

$= \left(\frac{1}{2} + \frac{1}{4} + 2\right) - (0)$

$= \frac{11}{4}$

**c)** Hence find the area of the shaded region.

Area of shaded region = area of trapezium − area under curve

To find the area of the trapezium, work out the values of $a$, $b$ and $h$:

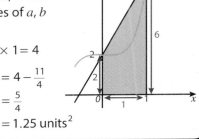

$A = \frac{1}{2}(2 + 6) \times 1 = 4$

Shaded area $= 4 - \frac{11}{4}$

$= \frac{5}{4}$

$= 1.25$ units$^2$

More challenging questions ask for the area between two curves, ask for an area which requires integration with respect to $y$, use algebraic terms, etc. The basic approach is just the same.

### Example

The graph shows the functions $f(y) = 3y^4 + y^3$ and $g(y) = 3y^4 + 5y^3 - 2y^2$.

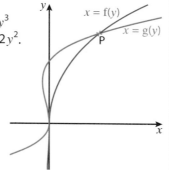

**a)** Find the coordinates of the point P.

To find the place where the graphs intersect,
$$3y^4 + y^3 = 3y^4 + 5y^3 - 2y^2$$
$$4y^3 - 2y^2 = 0$$
$$2y^2(2y - 1) = 0$$
$$y = 0, \ y = \frac{1}{2}$$
$$y = \frac{1}{2} \text{ as the graph shows } P > 0.$$
$$x = 3\left(\frac{1}{2}\right)^4 + \left(\frac{1}{2}\right)^3 = \frac{5}{16} = 0.3125$$
$$P = \left(\frac{5}{16}, \frac{1}{2}\right)$$

**b)** Find the area of the shaded region bounded by the curves and the $y$-axis.

Consider the area split into two parts:

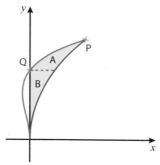

$Q$ is a solution to $3y^4 + 5y^3 - 2y^2 = 0$
$$y^2(3y^2 + 5y - 2) = 0$$
$$y^2(3y - 1)(y + 2) = 0$$
$$y = 0, \ y = \frac{1}{3}, \ y = -2$$

By inspection the point Q is $\left(0, \frac{1}{3}\right)$

Area A is $\int_{\frac{1}{3}}^{\frac{1}{2}} f(y) - g(y) \, dy = \int_{\frac{1}{3}}^{\frac{1}{2}} 2y^2 - 4y^3 \, dy$

$$= \left[\frac{2}{3}y^3 - y^4\right]_{\frac{1}{3}}^{\frac{1}{2}}$$

$$= \left(\frac{2}{3}\left(\frac{1}{2}\right)^3 - \left(\frac{1}{2}\right)^4\right) - \left(\frac{2}{3}\left(\frac{1}{3}\right)^3 - \left(\frac{1}{3}\right)^4\right)$$

$$= \frac{1}{48} - \left(\frac{1}{81}\right) = \frac{11}{1296}$$

Area B is $\int_0^{\frac{1}{3}} f(y) \, dy = \int_0^{\frac{1}{3}} 3y^4 + y^3 \, dy$

$$= \left[\frac{3}{5}y^5 + \frac{1}{4}y^4\right]_0^{\frac{1}{3}}$$

$$= \left(\frac{3}{5}\left(\frac{1}{3}\right)^5 + \frac{1}{4}\left(\frac{1}{3}\right)^4\right) - \left(\frac{3}{5}(0)^5 - \frac{1}{4}(0)^4\right)$$

$$= \frac{1}{180} - 0 = \frac{1}{180}$$

Total area $= A + B = \frac{11}{1296} + \frac{1}{180} = \frac{91}{6480}$ units$^2$

## Context

In context, questions might not mention area at all. Instead they may refer to displacement (e.g. a velocity–time graph, where the area under the curve is the displacement). The modulus of displacement is the distance travelled; in this case, the positive value is taken. If asked for displacement, keep the negative value if there is one.

### Example

Find the distance travelled by a car moving with velocity $v = t^2 - 6t + 10$ ms$^{-1}$, between times $t = 2$ s and $t = 6$ s.

$$\int_2^6 t^2 - 6t + 10 \, dx = \left[\frac{1}{3}t^3 - 3t^2 + 10t\right]_2^6$$

$$= \left(\frac{1}{3} \times 6^3 - 3 \times 6^2 + 10 \times 6\right)$$

$$- \left(\frac{1}{3} \times 2^3 - 3 \times 2^2 + 10 \times 2\right)$$

$$= 24 - \frac{32}{3} = \frac{40}{3}$$

Distance travelled = 13.33 metres (2 d.p.).

## Links to Other Concepts

- Polynomials ● Algebraic manipulation
- Indices ● Mechanics ● Differentiation
- Binomial expansion ● Rates of change

## SUMMARY

- $\int x^n \, dx = \dfrac{x^{n+1}}{n+1} + C, \; n \neq -1$

  **To integrate, add 1 to the power and divide the coefficient by the new power.**

- $\int ax^n + bx^m \, dx = a \int x^n \, dx + b \int x^m \, dx + C,$
  $n \neq -1, \; m \neq -1$

- **The '+ C' needs to be added to define the curve fully. Don't forget it!**

- **Key functions to know and recognise are:**

  $\int e^{kx} \, dx = \dfrac{1}{k} e^{kx} + C$   $\qquad \int \dfrac{1}{x} \, dx = \ln x + C$

  $\int \sin kx \, dx = -\dfrac{1}{k} \cos kx + C$

  $\int \cos kx \, dx = \dfrac{1}{k} \sin kx + C$

- **Use given values for the curve to find $C$.**

- **Integration can be used to find the area between a curve and the $x$-axis.**

- $A = \displaystyle\int_a^b f(x) \, dx$ **means the integral between $b$ and $a$ of the function $f(x)$ with respect to $x$. It finds the area between the curve $y = f(x)$ and the $x$-axis.**

- **If the area is below the $x$-axis, the integral between bounds will return a negative answer.**

- **If the integration between limits gives a negative answer, the area is below the $x$-axis. If asked for the area, take the modulus.**

- **Composite areas could be made up using a combination of curves and straight lines. At least one area will be the integral between limits and another could be the area of a right-angled triangle; the area of a trapezium; the area of a rectangle; or a second integral between limits.**

- **Make sure the final answer is in context.**

## QUICK TEST

1. Find the area bounded by the following, where $x$ is measured in radians.

   a) $y = \dfrac{360}{x} + \sin 2x$, the $x$-axis, $x = 200$ and $x = 450$

   b) $y = 2x$ and $y = (x - 3)(1 - 2x)$

2. Evaluate $\displaystyle\int 10e^{4x} - 5\cos\dfrac{x}{3} + \dfrac{3}{\sqrt[3]{x}} \, dx$

## PRACTICE QUESTIONS

1. You are given that $f'(x) = a \sin 2x + 2 \cos ax$, where $a$ is a constant.

   a) Find an expression for $f(x)$ in terms of $a$. **[3 marks]**

   The graph $y = f(x)$ passes through a point $\left(\dfrac{\pi}{12}, \dfrac{\sqrt{2}}{3}\right)$.

   b) Find an expression for $C$ in terms of $a$. **[3 marks]**

   The graph $y = f(x)$ also passes through the point $\left(-\dfrac{\pi}{2}, \left(\dfrac{3a+4}{6} + C\right)\right)$ and $a$ is positive, odd and an integer.

   c) What is the difference between the $y$-intercept for the graphs $y = f'(x)$ and $y = f(x)$? **[8 marks]**

2. The graph shows the curve $x = f(y)$ and $x = f'(y)$.

   $f(y) = -8\cos\left(\dfrac{y}{2}\right) + 4\sin\left(\dfrac{y}{2}\right) - 5$

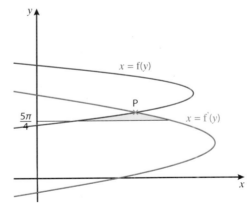

   a) Find $f'(y)$. **[2 marks]**

   b) Find the exact coordinate of point P. **[4 marks]**

   c) Hence find the shaded area bounded by the two curves and the line $y = \dfrac{5\pi}{4}$, as shown on the diagram. **[8 marks]**

# Integration 2

## Limit of Summation

The area underneath a curve can be estimated in a number of ways (e.g. using the trapezium rule). One method is to consider the area as being split into $n$ rectangles of width $\delta x$. This method of approximation leads to the formula:

$$\int_a^b y \, dx \approx \{y_1 + y_2 + \ldots + y_n\}\delta x$$

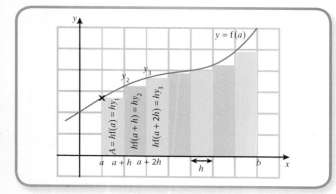

To get a more accurate approximation, the value of $n$ is increased, and hence the value of $\delta x$ decreased since they are inversely proportional: $\delta x = \frac{b-a}{n}$

As $n \to \infty$, $\quad \delta x \to 0$

**Area** $= \displaystyle\lim_{n \to \infty} \sum_{i=1}^{n} y_i \delta x$

The limit of the sum of the rectangles is equal to the definite integral.

$$\lim_{n \to \infty} \sum_{i=1}^{n} y_i \delta x = \int_a^b y \, dx$$

If a graph is increasing, then the approximation will be an underestimate.

If the graph is decreasing, then the approximation will be an overestimate.

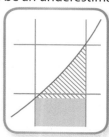

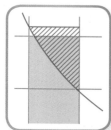

### Example

Erin estimates the area beneath the graph $y = \frac{x^3}{10} + \frac{x}{5} + 3$ between the lines $x = 3$ and $x = 5$ by calculating the area of four rectangles of equal width, $\delta x$.

**a)** What is her value of $\delta x$?

$$\delta x = \frac{b-a}{n}$$
$$= \frac{5-3}{4} = 0.5$$

**b)** Find the value of Erin's approximation.

$$\int_3^5 y \, dx \approx \{y_1 + y_2 + y_3 + y_4\}\delta x$$

| $x_1$ | $x_2$ | $x_3$ | $x_4$ |
|---|---|---|---|
| 3 | 3.5 | 4 | 4.5 |
| $y_1$ | $y_2$ | $y_3$ | $y_4$ |
| $\frac{3^3}{10} + \frac{3}{5} + 3$ $= 6.3$ | $\frac{3.5^3}{10} + \frac{3.5}{5} + 3$ $= 7.9875$ | $\frac{4^3}{10} + \frac{4}{5} + 3$ $= 10.2$ | $\frac{4.5^3}{10} + \frac{4.5}{5} + 3$ $= 13.0125$ |

$$\int_3^5 y \, dx \approx \{6.3 + 7.9875 + 10.2 + 13.0125\}0.5$$
$$\approx 37.5 \times 0.5$$
$$\approx 18.75$$

**c)** Erin says that $\frac{dy}{dx} = \frac{3x^2}{10} + \frac{1}{5}$, hence her estimate is an overestimate. Is she correct? Fully justify your answer.

The function $\frac{3x^2}{10} + \frac{1}{5}$ is always positive since $x^2 \geqslant 0$.

As $\frac{dy}{dx} > 0$ in the range $3 \leqslant x \leqslant 5$, the graph is increasing in this range.

As the graph is increasing, this method of approximation gives an underestimate, so Erin is incorrect.

**d)** Calculate the exact area under the curve and calculate the percentage error in Erin's approximation, showing your working.

$$\int_3^5 y\,dx = \left[\frac{x^4}{40} + \frac{x^2}{10} + 3x\right]_3^5$$

$$= \left(\frac{5^4}{40} + \frac{5^2}{10} + 3\times 5\right) - \left(\frac{3^4}{40} + \frac{3^2}{10} + 3\times 3\right)$$

$$= 33.125 - 11.925 = 21.2$$

$$\% \text{ error} = \frac{21.2 - 18.75}{21.2} \times 100 = 11.55660...$$

$$= 11.6\% \text{ (1 d.p.)}$$

# Integration by Substitution

Integration by substitution is used when confronted with a function in the form $y = f(g(x))$.

Let $g(x) = u$

$$\frac{du}{dx} = g'(x)$$

$$dx = \frac{du}{g'(x)}$$

$$\int f(g(x))\,dx = \int f(u)\frac{du}{g'(x)}$$

When $g(x)$ is a linear function, $g'(x)$ is a constant ($C$), so the integration becomes $\int f(g(x))\,dx = \frac{1}{C}\int f(u)\,du$.

**Example**
Use integration by substitution to find the integral of $y = 2\sec^2(4x+2)$.

Let $u = 4x + 2$

$$\frac{du}{dx} = 4$$

$$dx = \frac{du}{4}$$

$$\int y\,dx = \int 2\sec^2(u)\frac{du}{4} = \frac{1}{2}\int \sec^2(u)\,du$$

$$= \frac{1}{2}\tan(u) + C = \frac{1}{2}\tan(4x+2) + C$$

Sometimes $g(x)$ is not a linear function. In these cases, the function $g'(x)$ will in part simplify with the $f(x)$ function in order to leave an integral $\int f(g(x))\,dx = k\int h(u)\,du$, where $h(u)$ can be integrated with respect to $u$. If integrating between limits, calculate the new limits of integration in terms of $u$, rather than converting back to $x$.

Some questions will give the substitution, whilst others will ask for it to be used but you will need to select the part of the function to be substituted. Other questions will ask for an integral without suggesting substitution at all. If a function feels impossible to integrate in the current form, consider if a substitution could simplify the problem.

**Example**
Evaluate $\int_2^5 \frac{6x^2 - 2}{(x^3 - x + 5)^3}\,dx$ .

Let $u = x^3 - x + 5$

Limits of integration: $2^3 - 2 + 5 = 11$ and $5^3 - 5 + 5 = 125$

$$\frac{du}{dx} = 3x^2 - 1 \quad \rightarrow \quad dx = \frac{1}{3x^2 - 1}\,du$$

$$\int_2^5 \frac{6x^2 - 2}{(x^3 - x + 5)^3}\,dx = \int_2^5 \frac{2\left(\frac{du}{dx}\right)}{(u)^3}\,dx$$

$$= \int_{11}^{125} 2u^{-3}\,du$$

$$= \left[-u^{-2}\right]_{11}^{125}$$

$$= (-125^{-2}) - (-11^{-2}) = 0.00820 \text{ (3 s.f.)}$$

**Expressions of the Form** $\int \frac{f'(x)}{f(x)}\,dx$

Let $u = f(x) \quad \rightarrow \quad \frac{du}{dx} = f'(x) \quad \rightarrow \quad dx = \frac{du}{f'(x)}$

$$\int \frac{f'(x)}{f(x)}\,dx \quad \rightarrow \quad \int \frac{f'(x)}{u}\frac{du}{f'(x)} \quad \rightarrow \quad \int \frac{1}{u}\,du$$

$$\int \frac{f'(x)}{f(x)}\,dx \quad \rightarrow \quad \ln(f(x)) + C$$

**Example**
$y = \frac{\sin x}{\cos x}$, find $\int y\,dx$.

Remember: Differentiating
Integrating

$$-\int -\frac{\sin x}{\cos x}\,dx = -\ln(\cos x) + C$$

## Expressions of the Form $\int f'(x)\big[f(x)\big]^n \, dx$

Let $u = f(x)$

$\frac{du}{dx} = f'(x)$

$\int f'(x)[f(x)]^n \, dx = \int \frac{du}{dx} [u]^n \, dx = \int [u]^n \, du$

---

### Example

Integrate $2\sin(4x)\cos^4(4x)$.

Since $\frac{d}{dx}(\cos 4x) = -4\sin 4x$

and $2\sin 4x = -\frac{1}{2}(-4\sin 4x)$

$\int 2\sin(4x)\cos^4(4x)\,dx = -\frac{1}{2}\int u^4 \, du$

$\qquad = -\frac{1}{2} \times \frac{1}{5}u^5 + C$

$\qquad = -\frac{1}{10}\cos^5(4x) + C$

---

### Don't Forget

When integrating $\frac{dv}{dx}$ to get an expression for $v$, the constant of integration isn't included as it gets cancelled out.

$\int u\frac{dv}{dx}\,dx = u(v+C) - \int (v+C)\frac{du}{dx}\,dx$

$\qquad = uv + uC - \int (v)\frac{du}{dx}\,dx - \int C\frac{du}{dx}\,dx$

$\qquad = uv - \int (v)\frac{du}{dx}\,dx + uC - uC$

Instead of showing all the steps, it is standard practice to omit the constant of integration.

---

## Integration by Parts

Integration by parts is used when there are two functions multiplied together, $u$ and $\frac{dv}{dx}$.

The product rule for differentiation says that:

$\frac{d(uv)}{dx} = v\frac{du}{dx} + u\frac{dv}{dx}$

Integrating with respect to $x$ gives:

$uv = \int v\frac{du}{dx}\,dx + \int u\frac{dv}{dx}\,dx$

Rearranging gives:

$\int u\frac{dv}{dx}\,dx = uv - \int v\frac{du}{dx}\,dx$

In each case, decide which function should be treated as $u$ and which as $\frac{dv}{dx}$.

$\frac{dv}{dx}$ needs to be integrated to find $v$, so needs to be possible to integrate.

---

$u$ needs to be differentiated to get $\frac{du}{dx}$. If there is one part that is a polynomial, it will generally be $u$.

The product $v\frac{du}{dx}$ needs to be integrated; this can sometimes take more than one application of integration by parts.

---

### Example

Find $\int \frac{1}{2}x\ln x\,dx$.

The constant can be taken to the front of the integral $\left(\text{i.e. } \frac{1}{2}\int x\ln x\,dx\right)$ or be included as part of either $u$ or $\frac{dv}{dx}$.

Let $u = \ln x \quad \rightarrow \quad \frac{du}{dx} = \frac{1}{x}$

(since $\ln x$ cannot be easily integrated, this should be $u$ rather than $\frac{dv}{dx}$)

Let $\frac{dv}{dx} = \frac{1}{2}x \quad \rightarrow \quad v = \frac{1}{4}x^2$

$\int \frac{1}{2}x\ln x\,dx = \frac{1}{4}x^2\ln x - \int \frac{1}{4}x^2 \times \frac{1}{x}\,dx$

$\qquad = \frac{1}{4}x^2\ln x - \int \frac{x}{4}\,dx$

$\qquad = \frac{1}{4}x^2\ln x - \frac{x^2}{8} + C$

---

### Example

Find $\int (x^2 + 2x)\cos x\,dx$.

Let $u = x^2 + 2x \qquad \rightarrow \quad \frac{du}{dx} = 2x + 2$

Let $\frac{dv}{dx} = \cos x \qquad \rightarrow \quad v = \sin x$

$\int (x^2+2x)\cos x\,dx = (x^2+2x)\sin x - \int (2x+2)\sin x\,dx$

This is still in a form which needs integration by parts to be applied.

Let $u = 2x + 2 \qquad \rightarrow \quad \frac{du}{dx} = 2$

Let $\frac{dv}{dx} = \sin x \qquad \rightarrow \quad v = -\cos x$

$(x^2+2x)\sin x - \int (2x+2)\sin x\,dx$

$= (x^2+2x)\sin x - \left[(2x+2)(-\cos(x)) - \int -2\cos x\,dx\right]$

$= (x^2+2x)\sin x - [-(2x+2)\cos x + 2\sin x] + C$

$= (x^2+2x)\sin x + (2x+2)\cos x - 2\sin x + C$

$= (x^2+2x-2)\sin x + 2(x+1)(\cos x) + C$

One of the common mistakes for a question like the previous one occurs when it comes to dealing with the negatives. Take your time and show your working clearly to avoid mistakes. Another common mistake is to miss the constant $(+C)$, which should be present once the final part of the integration is completed.

Keep an eye out for expressions that can be simplified.

---

**Example**

$\frac{dy}{dx} = \frac{12x^3 \sin(x + \pi)}{3x^2}$

Find the value of $y$ at $x = -\pi$ given that at $x = \pi$, $y = -\pi$.

Let $u = \left(\frac{12x^3}{3x^2}\right) = 4x$ $\qquad \rightarrow \frac{du}{dx} = 4$

Let $\frac{dv}{dx} = \sin(x + \pi) = -\sin(x)$ $\rightarrow$ $v = \cos x$

$y = 4x \cos x - \int 4 \cos x \, dx$

$\quad = 4x \cos x - 4 \sin x + C$

$-\pi = 4\pi(-1) - 4(0) + C$

$C = -\pi + 4\pi = 3\pi$

$y = 4x \cos x - 4 \sin x + 3\pi$

At $x = -\pi$

$y = 4(-\pi)(-1) - 4(0) + 3\pi = 7\pi$

---

## Integration by Inspection

There are times when a full method of integration can be skipped as they occur so often, or can be easily recognised from 'what would differentiate to give the expression'. These include:

- $\int e^{ax+b} \, dx = \frac{1}{a} e^{ax+b} + C$

- $\int \sin(ax+b) \, dx = -\frac{1}{a} \cos(ax+b) + C$ and

  $\int \cos(ax+b) \, dx = \frac{1}{a} \sin(ax+b) + C$

- $\int \frac{1}{ax+b} \, dx = \frac{1}{a} \ln(ax+b) + C$

- Expressions of the form $\int \frac{f'(x)}{f(x)} \, dx$ and

  $\int f'(x)[f(x)]^n \, dx$.

---

For each of these, substitution or integration by parts could be used. If unsure, use the full method to solve.

## Partial Fractions

As met on page 14, partial fractions are used to change the form of an algebraic fraction. They can be used to simplify a fraction into a form that can be more easily integrated. The requirement to use partial fractions could be made explicitly as part of a question or it may not be mentioned at all. If there is a question with polynomial expressions in both the numerator and the denominator, consider whether a partial fraction might be a good way to approach integration.

---

**Example**

Find an expression for the integral of

$y = \frac{6x^2 + 3x - 2}{(x - 2)(x + 1)(x + 3)}$

Using partial fractions:

$\frac{6x^2 + 3x - 2}{(x - 2)(x + 1)(x + 3)} = \frac{A}{x - 2} + \frac{B}{x + 1} + \frac{C}{x + 3}$

$(6x^2 + 3x - 2) = A(x + 1)(x + 3) + B(x - 2)(x + 3)$

$+ C(x - 2)(x + 1)$

When $x = -1$

$1 = B(-3)(2) = -6B$ $\qquad \rightarrow$ $B = -\frac{1}{6}$

When $x = -3$

$43 = C(-5)(-2) = 10C$ $\qquad \rightarrow$ $C = \frac{43}{10}$

When $x = 2$

$28 = A(3)(5) = 15A$ $\qquad \rightarrow$ $A = \frac{28}{15}$

$\int y \, dx = \int \frac{28}{15(x - 2)} - \frac{1}{6(x + 1)} + \frac{43}{10(x + 3)} \, dx$

$= \frac{28}{15} \ln(x - 2) - \frac{1}{6} \ln(x + 1) + \frac{43}{10} \ln(x + 3) + C$

---

**Links to Other Concepts**
- Graph sketching ● Polynomials
- Algebraic manipulation ● Indices
- Differentiation ● Problem solving
- Rates of change
- Natural logs and exponentials
- Trigonometry ● Differential equations
- Mechanics and rates of change

## SUMMARY

- Other elements that could be included in an integration question are:
  - manipulating expressions (including logs and trigonometric functions)
  - simplifying expressions (factor theorem, etc.)
  - parametric equations, or equations in terms of different variables
  - finding roots of equations to define limits
  - finding the value of the constant of integration
  - finding the area bounded by a curve (or by two curves)
  - finding the value of the integrated function
  - context-based elements
  - finding definite integrals.

- $\lim\limits_{n \to \infty} \sum\limits_{i=1}^{n} y_i \delta x = \int_a^b y \, dx$
  - $\delta x = \dfrac{b-a}{n}$
  - Increasing graph $\left(\dfrac{dy}{dx} > 0\right)$ gives an underestimate
  - Decreasing graph $\left(\dfrac{dy}{dx} < 0\right)$ gives an overestimate.

- Substitution is used when there is a function within a function:
  - Let $u$ equal part of the function and find $\dfrac{du}{dx}$
  - Substitute $u$ and $dx$ into the original function and simplify
  - Integrate with respect to $u$ and either convert back to $x$ or use adjusted limits to find a definite integral.

- Integration by parts is used when integrating the product of two functions:
  - $\int u \dfrac{dv}{dx} \, dx = uv - \int v \dfrac{du}{dx} \, dx$
  - May need more than one application to fully integrate.

- Remember the $+C$.

- Partial fractions can be used to make an expression easier to integrate.

## QUICK TEST

1.  Find the integral of $y$ with respect to $x$ for each of these.

    a)  $y = e^{3x-2}$

    b)  $y = 56 \cos(7x - 3)$

    c)  $y = \dfrac{1}{9x - 1}$

    d)  $y = \dfrac{1}{e^{2x-1}}$

2.  Write the following expressions in a form that can be integrated.

    a)  $\dfrac{3x^2 - 6x - 9}{(x + 1)(x - 3)(x + 2)}$

    b)  $\dfrac{x^2 + 2x - 12}{x(x^2 - 4)}$

3.  Use the substitution given in each case to find an expression for the integral.

    a)  $\displaystyle\int \dfrac{2}{(2x - 5)^{\frac{1}{2}}} \, dx, \ u = 2x - 5$

    b)  $\displaystyle\int \cos(3x) e^{\sin(3x)} \, dx, \ u = \sin 3x$

    c)  $\displaystyle\int \dfrac{3 \ln 2x}{x} \, dx, \ u = \ln(2x)$

4.  $\dfrac{dy}{dx} = \sin^2\left(2x - \dfrac{\pi}{2}\right)$

    a)  Use $\sin^2 A = \dfrac{1}{2}(1 - \cos 2A)$ to find an expression for $y$ in terms of $x$.

    b)  Find the value of $y$ when $x = \dfrac{\pi}{4}$, given that at $x = \dfrac{\pi}{8}$, $y = \dfrac{\pi}{16}$.

5.  In each of these, decide which part of the function should be $u$ and which $\dfrac{dv}{dx}$.

    a)  $f(x) = (4x + 2)\sin 2x$

    b)  $f(x) = e^x 2x$

    c)  $f(x) = x\sqrt{x + 3}$

    d)  $f(x) = (3x^2 + 2x - 1)(x - 3)^5$

6.  Find $\dfrac{du}{dx}$ and $v$ for each function from question 5.

7.  Fully integrate the functions from question 5.

# PRACTICE QUESTIONS

**1.** $g(x) = \dfrac{2x(x+2)}{(x-1)(x^2+2x-3)} + 1$

Find the integral of $g(x)$. **[7 marks]**

**2.** $f(\theta) = \theta^2 \cos\left(\theta + \dfrac{\pi}{4}\right)$ and $\displaystyle\int g(\theta)\,d\theta = \dfrac{2}{27}\cos 3\theta + \dfrac{2\theta}{9}\sin 3\theta - \dfrac{\theta^2}{3}\cos 3\theta$

**a)** Find $\displaystyle\int f(\theta)\,d\theta$. **[6 marks]**

**b)** Confirm that the points of intersection of $f(\theta)$ and $g(\theta)$ lie at $\theta_P = -\dfrac{5\pi}{8}, \theta_Q = -\dfrac{7\pi}{16}$ and $\theta_R = 0$. Clearly show your working and reasoning. **[8 marks]**

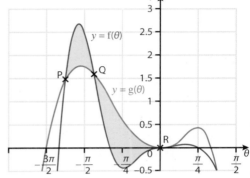

**c)** Which of the following are correct expressions to find the area of the shaded region? **[2 marks]**

| | |
|---|---|
| **A** | $\left\| \displaystyle\int_{-\frac{5\pi}{8}}^{0} f(\theta)\,d\theta - \int_{-\frac{5\pi}{8}}^{0} g(\theta)\,d\theta \right\|$ |
| **B** | $\left\| \displaystyle\int_{-\frac{7\pi}{16}}^{0} f(\theta) - g(\theta)\,d\theta \right\| + \left\| \int_{-\frac{5\pi}{8}}^{-\frac{7\pi}{16}} f(\theta) - g(\theta)\,d\theta \right\|$ |
| **C** | $\left\| \displaystyle\int_{-\frac{7\pi}{16}}^{0} f(\theta) - g(\theta)\,d\theta + \int_{-\frac{5\pi}{8}}^{-\frac{7\pi}{16}} f(\theta) - g(\theta)\,d\theta \right\|$ |
| **D** | $\left\| \displaystyle\int_{-\frac{7\pi}{16}}^{0} f(\theta) - g(\theta)\,d\theta \right\| - \left\| \int_{-\frac{5\pi}{8}}^{-\frac{7\pi}{16}} g(\theta) - f(\theta)\,d\theta \right\|$ |
| **E** | $\left\| \displaystyle\int_{-\frac{7\pi}{16}}^{0} f(\theta) - g(\theta)\,d\theta - \int_{-\frac{5\pi}{8}}^{-\frac{7\pi}{16}} f(\theta) - g(\theta)\,d\theta \right\|$ |

**3.** The graph shows the curve $x = 2\sin\theta$, $y = \sin\theta\,\cos^5\theta$, $0 \leqslant \theta \leqslant 2\pi$.

Find the area of the shaded region. **[10 marks]**

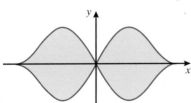

# Differential Equations

A differential equation is an equation that contains derivatives, i.e. $\frac{dy}{dx}$. This could be presented as a question with a term including $\frac{dy}{dx}$ or $\frac{dx}{dy}$ but it could also be given as 'the gradient of a curve at any given point' being equal to an expression in terms of both $x$ and $y$. To solve differential equations, manipulating expressions is important in order to separate the variables and to get the equation into the required form.

## First Order Differential Equations

The highest power of the derivatives in the equation is 1. At A-level, the differential equations will be simple, which means they have variables that can be easily separated and are first order.

The general solution to the equation $\frac{dy}{dx} = f(x)g(y)$ is:

$$\int \frac{1}{g(y)}\,dy = \int f(x)\,dx + C$$

A particular solution can be found if given a point $(x, y)$ in order to find the constant of integration ($C$).

---

**Example**

**a)** Find the general solution for the differential equation $2\frac{dy}{dx} = \frac{x^2 + 3x}{4y - 1}$

Multiply both sides by $(4y - 1)$:

$(8y - 2)\frac{dy}{dx} = x^2 + 3x$

Separate the variables and integrate:

$$\int (8y - 2)\,dy = \int (x^2 + 3x)\,dx$$

$4y^2 - 2y = \frac{1}{3}x^3 + \frac{3}{2}x^2 + C$

**b)** Find the particular solution given that the graph passes through the point $(0, -0.5)$.

$4(-0.5)^2 - 2(-0.5) = 0 + C$

$C = 2$

The particular solution is:

$4y^2 - 2y = \frac{1}{3}x^3 + \frac{3}{2}x^2 + 2$

---

**c)** Hence find the other point that sits on the $y$-axis.

$4y^2 - 2y = 2$

$4y^2 - 2y - 2 = 0$

$2y^2 - y - 1 = 0$

$(2y + 1)(y - 1) = 0$

Solutions $-0.5$ and $1$

Second point is $(0, 1)$.

---

Sometimes the variables are easy to split into a function of $x$ and a function of $y$. At other times they are more complex. Any of the methods of integration could be needed for each function.

---

**Example**

$2\tan y \frac{dy}{dx} = \frac{x(1 - \tan^2 y)}{(2x + 3)(x - 4)}$

Show that $\sec(2y) = (2x + 3)^{\frac{3}{11}}(x - 4)^{\frac{8}{11}}C$, where $C$ is a constant.

$\frac{2\tan y}{1 - \tan^2 y}\frac{dy}{dx} = \frac{x}{(2x + 3)(x - 4)}$

$\int \frac{2\tan y}{1 - \tan^2 y}\,dy = \int \frac{x}{(2x + 3)(x - 4)}\,dx$

The variables are now separated, so each side needs to be integrated.

$\int \frac{2\tan y}{1 - \tan^2 y}\,dy = \int \tan 2y\,dy$    (using double angle formulae)

Given that $\frac{d}{dy}(\cos 2y) = -2\sin(2y)$

$\int \tan 2y\,dy = -\frac{1}{2}\int -\frac{2\sin 2y}{\cos 2y}\,dy = -\frac{1}{2}\ln|\cos 2y| + k_1$

$= \ln\sqrt{\sec 2y} + k_1$

$\int \frac{x}{(2x + 3)(x - 4)}\,dx$

Using partial fractions:

$\frac{x}{(2x + 3)(x - 4)} = \frac{A}{2x + 3} + \frac{B}{x - 4}$

$x = A(x - 4) + B(2x + 3)$

---

When $x = 4$

$4 = 11B \rightarrow \quad B = \dfrac{4}{11}$

When $x = -\dfrac{3}{2}$

$-\dfrac{3}{2} = -\dfrac{11}{2}A \quad \rightarrow \quad A = \dfrac{3}{11}$

$\displaystyle\int \dfrac{x}{(2x+3)(x-4)}\,dx = \int \dfrac{3}{11(2x+3)} + \dfrac{4}{11(x-4)}\,dx$

$\qquad\qquad\qquad = \dfrac{3}{22}\ln(2x+3) + \dfrac{4}{11}\ln(x-4) + k_2$

$\ln\sqrt{\sec 2y} = \dfrac{3}{22}\ln(2x+3) + \dfrac{4}{11}\ln(x-4) + \ln k_3$

$\qquad$ (where $\ln k_3 = k_2 - k_1$)

$\qquad\qquad = \ln(2x+3)^{\frac{3}{22}} + \ln(x-4)^{\frac{4}{11}} + \ln k_3$

$\qquad\qquad = \ln[(2x+3)^{\frac{3}{22}}(x-4)^{\frac{4}{11}}k_3]$

$e^{\ln\sqrt{\sec 2y}} = e^{\ln[(2x+3)^{\frac{3}{22}}(x-4)^{\frac{4}{11}}k_3]}$

$\sqrt{\sec 2y} = (2x+3)^{\frac{3}{22}}(x-4)^{\frac{4}{11}}k_3 \quad$ Let $C = (k_3)^2$

$\sec(2y) = (2x+3)^{\frac{3}{11}}(x-4)^{\frac{8}{11}}C$

$\dfrac{y}{x}$ can appear in more than one term, in which case gather like terms:

## Example

$\dfrac{1}{x} \times \dfrac{dx}{dy} + 4y^3 = 2x \times \dfrac{dx}{dy}$

$4y^3 = 2x \times \dfrac{dx}{dy} - \dfrac{1}{x} \times \dfrac{dx}{dy}$

$\qquad = \left(2x - x^{-1}\right)\dfrac{dx}{dy}$

$\displaystyle\int 4y^3\,dy = \int 2x - x^{-1}\,dx$

$y^4 = x^2 - \ln x + C$

## Context

Differential equations occur in 'real-life situations', so you need to be able to model context-based questions as differential equations in order to solve them. Many questions relate to rates of change (differentiation and gradient). It is important to be familiar with all the formulae for areas and volumes from GCSE, as well as spotting differential equations in the context of mechanics, depreciation, growth, etc.

## Example

A student models the growth of premature babies such that the rate of change of their mass ($m$) with respect to time ($t$) is inversely proportional to the product of the square root of their mass (in kg) and the sum of a tenth of their age (in days) add 2.

a) Write a differential equation for the relationship in terms of $m$, $t$ and a constant $k$.

$\dfrac{dm}{dt} = \dfrac{k}{\sqrt{m}\left(\dfrac{t}{10}+2\right)}$

b) It is given that $k = \dfrac{7}{30}$.

At birth, a baby had a mass of 1.69 kg.

Using the suggested model, how many days old would you expect the baby to be when it first weighs over 2.5 kg?

$\dfrac{dm}{dt} = \dfrac{7}{30m^{\frac{1}{2}}\left(\dfrac{t}{10}+2\right)}$

$\displaystyle\int \dfrac{30}{7}m^{\frac{1}{2}}\,dm = \int \dfrac{1}{\left(\dfrac{t}{10}+2\right)}\,dt$

$\dfrac{30}{7} \times \dfrac{2}{3}m^{\frac{3}{2}} = 10\ln\left(\dfrac{t}{10}+2\right) + C_1$

$m^{\frac{3}{2}} = \dfrac{7}{2}\ln\left(\dfrac{t}{10}+2\right) + C$

At $t = 0$, $m = 1.69$

$1.69^{\frac{3}{2}} = \dfrac{7}{2}\ln(2) + C$

$C = 2.197 - \dfrac{7}{2}\ln 2$

$m^{\frac{3}{2}} = \dfrac{7}{2}\ln\left(\dfrac{t}{10}+2\right) + 2.197 - \dfrac{7}{2}\ln 2$

Find $t$ when $m = 2.5$:

$2.5^{\frac{3}{2}} = \dfrac{7}{2}\ln\left(\dfrac{t}{10}+2\right) + 2.197 - \dfrac{7}{2}\ln 2$

$2.5^{\frac{3}{2}} - 2.197 + \dfrac{7}{2}\ln 2 = \ln\left(\dfrac{t}{10}+2\right)^{\frac{7}{2}}$

$\left(\dfrac{t}{10}+2\right)^{\frac{7}{2}} = e^{2.5^{\frac{3}{2}} - 2.197 + \frac{7}{2}\ln 2}$

$\dfrac{t}{10}+2 = e^{\frac{2}{7}(2.5^{\frac{3}{2}} - 2.197 + \frac{7}{2}\ln 2)} = 3.302955\ldots$

$t = (3.302955\ldots - 2) \times 10 = 13.02955\ldots$

You would expect the baby to be 13 days old.

# DAY 5

## Links to Other Concepts

● Calculus ● Exponentials ● Polynomials ● Algebraic manipulation ● Indices ● Mechanics
● Problem solving ● Rates of change ● Trigonometry ● Context-based questions
● Geometry: properties of shapes

## SUMMARY

● Differential equations can be manipulated such that an equation in terms of $x$, $y$ and $\frac{dy}{dx}$ can be written as integrals of $f(x)$ and $g(y)$.

● Equations may be given or may need to be formed using context-based information.

● To get a general solution to a differential equation $\frac{dy}{dx} = \frac{f(x)}{g(y)}$:

$$\int g(y)\,dy = \int f(x)\,dx + C$$

● To get a general solution to a differential equation $\frac{dy}{dx} = f(x)\,g(y)$:

$$\int \frac{1}{g(y)}\,dy = \int f(x)\,dx + C$$

● Always remember $+C$ when integrating.

● A particular solution can be found by using a pair of values $(x, y)$ to find the constant of integration ($C$).

**1.** Match the differential equation to its partial and full general solution. Fill in the missing ones.

| Differential equation | Partial solution | General solution |
|---|---|---|
| $\frac{dy}{dx} = 2y^4(x+3)$ | $\int 7y^{-4}\,dy = \int 5x^{-1}\,dx$ | $-\frac{1}{3y^3} = x^2 + 6x + C$ |
| $7x\frac{dy}{dx} - 5y^4 = 0$ | | |
| $\frac{dx}{dy} = 2y^4(x+3)$ | $\int 6y^2\,dy = \int 2x\sin x\,dx$ | $\frac{4}{3}y^3 = \ln x - 2x + C$ |
| $\frac{dx}{dy} - 4xy^2 = 2x\frac{dx}{dy}$ | $\int x\sin^2 x\,dx = \int y^{-3}\,dy$ | $\frac{2}{5}y^5 = \ln(x+3) + C$ |
| $\sin x\frac{dx}{dy} = \frac{1}{xy^3\sin x}$ | $\int y^{-4}\,dy = \int 2x + 6\,dx$ | $-\frac{7}{3y^3} = 5\ln x + C$ |
| $2x = \frac{\frac{dy}{dx}6y^2}{\sin x}$ | $\int x^{-1} - 2\,dx = \int 4y^2\,dy$ | $-\frac{1}{2y^2} = \frac{x^2}{4} - \frac{x}{4}\sin 2x + \frac{1}{8}\cos 2x + C$ |

**2.** For each of these general solutions, find the particular solution.

**a)** $2y^4 + 3y = x^2 + 2x + C$, $(5, 2)$

**b)** $\sin\left(y - \frac{\pi}{8}\right) = \frac{1}{2}x^2 + C$, $\left(\sqrt[4]{32}, \frac{3\pi}{8}\right)$

**c)** $\frac{2}{5}y^5 = \ln(x+3) + C$, $(-2, 1)$

**d)** $-\frac{7}{3y^3} = 2\ln x + C$, $(e^2, 3)$

**3.** Find the particular solutions to the following differential equations.

**a)** $\frac{dy}{dx} = 2xy + 3y$, $(0, 1)$

**b)** $\frac{dy}{dx} = e^{x-3y}$, $(0, 2)$

**c)** $\frac{dy}{dx} = \operatorname{cosec} y \cot x$, $\left(\frac{\pi}{2}, \frac{\pi}{4}\right)$

**d)** $\sqrt{y^3}\frac{dy}{dx} - 2y^7 = 0$, $\left(-1\frac{1}{9}, 1\right)$

**4.** Find the general solution to the differential equation.

$$x\frac{dy}{dx} - (8x+1)y^3 = 2\frac{dy}{dx} - 4x^2\frac{dy}{dx}$$

# PRACTICE QUESTIONS

**1. a)** $\sin x\frac{dy}{dx} = (\cos x - 4\sin x)\tan y$

Find the general solution. **[5 marks]**

**b)** Find the particular solution given that one pair of coordinates is $\left(\frac{2\pi}{3}, \frac{2\pi}{3}\right)$. **[3 marks]**

**2.** The resultant force applied to an object varies with time(s) and the velocity (ms$^{-1}$) of the object such that:

$$F = \frac{2\cos(2t - \pi) + 2t}{3v^2}$$

Given that the velocity of a body with mass 100 g, at a time 0 seconds, is −2 metres per second, what is the velocity at a time of 2 minutes? **[10 marks]**

# Numerical Methods of Approximation

## Locating Roots

You can rearrange any equation to be a function that is equal to zero. When the graph is plotted, the roots give a result to the equation, as previously met when solving quadratics. If $f(x) = g(x)$, then $f(x) - g(x) = 0$.

Often the first step in the numerical methods is to find a close estimate to where the root is. By showing a change of sign of function, a small interval in which the root sits can be identified. To do this, you need to be aware of the form of the graph (see page 18 about locating asymptotes). An asymptote represents a break in a curve and so a change in sign can represent passing an asymptote rather than passing a root.

---

**Example**

Explain why the change of sign between $x = 0.5$ and $x = 1.5$ for the function $f(x) = \frac{\cos(x^2)}{(x-1)(x+2)}$

doesn't give an interval containing a root to the equation $f(x) = 0$, but the change of sign between $x = -0.5$ and $x = -1.5$ does indicate the location of a root.

There are vertical asymptotes at $x = 1$ and $x = -2$. As the interval $0.5 < x < 1.5$ contains one of these asymptotes, it is not continuous in this region so the change of sign doesn't necessarily indicate the location of a root.

As neither vertical asymptote is located within the interval $-1.5 < x < -0.5$, the function is continuous in this range and the change of sign of the function indicates that there is at least one root in the interval.

---

If $f(a) = 0$, then $\alpha$ is a solution to the equation $f(x) = 0$. If the function is continuous, then to go from being positive to negative it must pass through 0.

If $f(a)$ is positive and $f(b)$ is negative, then there is a root of the equation between $a$ and $b$.

---

**Example**

a) Show that a solution to the equation $\sin(10x) = 3x^2 + 2x$ lies between 0.2 and 0.3, where $x$ is measured in radians.

$\sin(10x) - 3x^2 - 2x = 0$
Let $f(x) = \sin(10x) - 3x^2 - 2x$
$f(0.2) = \sin(2) - 3 \times 0.2^2 - 2 \times 0.2$
$\qquad = 0.38929...$
$f(0.3) = \sin(3) - 3 \times 0.3^2 - 2 \times 0.3$
$\qquad = -0.72887...$

The change of sign of a continuous function means that there is a root, $\alpha$, where $0.2 < \alpha < 0.3$.

b) Vanessa says that there are no roots between $-0.7$ and 0.3. Explain why she might think this and whether or not she is correct.

$f(-0.7) = \sin(-7) - 3(-0.7)^2 - 2(-0.7)$
$\qquad = -0.726986...$

The function is continuous and there is no change of sign; both $f(-0.7)$ and $f(0.3)$ give negative results. However, this might be because there are an even number of roots within the gap. Given that we know there is at least one root between 0.2 and 0.3 (from part **a**), we can conclude that there are an even number of roots between $-0.7$ and 0.3.

---

## Approximating Solutions to Equations

### Iteration

If an equation $y = f(x)$ has a solution $\alpha$, then you can set up an iterative formula to find increasingly accurate approximations for $\alpha$. The iterative formula will be of the form $x_{n+1} = g(x_n)$.

If the starting point is close enough to $\alpha$, then each iteration can produce a closer approximation. The sequence is said to converge on the root if, as $n$ increases

In the sequence $x_1, x_2, x_3, x_4, \ldots, x_n, \ldots$, the value of $x_n$ gets closer to the root $\alpha$. See page 43 for the use of recurrence relationships and calculators when finding terms of a sequence.

Some iterations with a given starting point will converge to a root, but others will diverge (move away) from the root.

**Staircase** and **cobweb diagrams** are used to represent the iterative approach. By plotting $y = x$ and $y = f(x)$, where $f(x)$ is the iterative formula, the roots are where the two lines cross. A graphical calculator can plot these diagrams for you. The iteration works by starting at $x_1$ and finding the associated value of $f(x)$, thus giving the next approximation, $x_2$. By moving from the function line horizontally, the $x$-value of the position now represents $x_2$. Moving vertically from $x_2$ will give a $y$-value of the function, which is the next approximation, $x_3$.

A staircase diagram is so-called because each approximation 'steps' closer to the root.

---

### Example
There is a root to the equation $4x - \sec^2 x = 0$ in the interval $[0.1, 0.8]$ (radians). Use the iterative relationship $x_{n+1} = \frac{1}{4}\sec^2 x$ to find the root, accurate to 4 decimal places within the interval.

In this case, either end of the given interval can be taken. As shown on the staircase diagram, both will converge on the root within the interval.

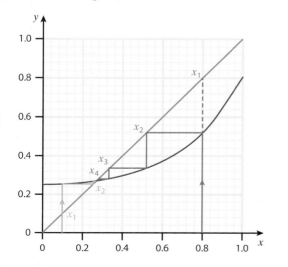

**Starting at 0.1**

$x_1 = 0.1$

$x_2 = \frac{1}{4}\sec^2 x_1 = 0.2525167\ldots$

$x_3 = \frac{1}{4}\sec^2 x_2 = 0.2666441\ldots$

$x_4 = \frac{1}{4}\sec^2 x_3 = 0.2686525\ldots$

$x_5 = \frac{1}{4}\sec^2 x_4 = 0.2689486\ldots \quad \rightarrow \quad 0.2690$ (4 d.p.)

$x_6 = \frac{1}{4}\sec^2 x_5 = 0.268992\ldots \quad \rightarrow \quad 0.2690$ (4 d.p.)

**Starting at 0.8**

$x_1 = 0.8$

$x_2 = \frac{1}{4}\sec^2 x_1 = 0.515038\ldots$

$x_3 = \frac{1}{4}\sec^2 x_2 = 0.330087\ldots$

$x_4 = \frac{1}{4}\sec^2 x_3 = 0.2793474\ldots$

$x_5 = \frac{1}{4}\sec^2 x_4 = 0.270570\ldots$

$x_6 = \frac{1}{4}\sec^2 x_5 = 0.2692338\ldots$

$x_7 = \frac{1}{4}\sec^2 x_6 = 0.2690348\ldots \quad \rightarrow \quad 0.2690$ (4 d.p.)

$x_8 = \frac{1}{4}\sec^2 x_7 = 0.2690052\ldots \quad \rightarrow \quad 0.2690$ (4 d.p.)

Since the last two values (in either method) give the same answer to 4 decimal places, you can test the interval for which this is the rounded answer to ensure that the root is as accurate as required.

$0.26895 < 0.2690 < 0.26905$

i.e. $f(0.26895) = 4 \times 0.26895 - \dfrac{1}{(\cos 0.26895)^2}$

$= -0.00017081\ldots$

$f(0.26905) = 4 \times 0.26905 - \dfrac{1}{(\cos 0.26905)^2}$

$= 0.00016985\ldots$

The change of sign of the continuous function confirms that a root lies within the interval, meaning that the root has a value 0.2690 accurate to 4 d.p.

An iterative formula doesn't always do what might be expected.

## Example

There is a second root for $4x - \sec^2 x = 0$ in the interval [0.9, 1.1] (radians). Angus hopes to find an approximation for this root using the same iterative formula as in the example on page 85. Will it work?

Using the same iterative formula, but starting at $x_1 = 1.1$, gives the following sequence:

$x_1 = 1.1$

$x_2 = \frac{1}{4}\sec^2 x_1 = 1.2150\ldots$

$x_3 = \frac{1}{4}\sec^2 x_2 = 2.0611\ldots$

$x_4 = \frac{1}{4}\sec^2 x_3 = 1.1273\ldots$

$x_5 = \frac{1}{4}\sec^2 x_4 = 1.3578\ldots$

$x_6 = \frac{1}{4}\sec^2 x_5 = 5.5997\ldots$

$x_7 = \frac{1}{4}\sec^2 x_6 = 0.4157\ldots$

$x_8 = \frac{1}{4}\sec^2 x_7 = 0.2987\ldots$

$x_9 = \frac{1}{4}\sec^2 x_8 = 0.2737\ldots$

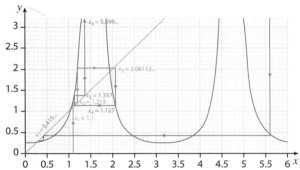

At this point, the iteration starts to converge on the root 0.2690 (4 d.p.). The diagram initially appears to diverge before converging on the root.

Convergence can happen from both sides of the root. As shown next, this produces a converging cobweb diagram.

## Example

The iterative formula $x_{n+1} = \sin(x_n + 5)$ is used with $x_1 = -1.5$ to find the root to the equation $\sin(x + 5) - x = 0$.

The sequence of approximations is as follows:

$-1.5, -0.35078\ldots, -0.99800\ldots, -0.75810\ldots,$
$-0.89134\ldots, -0.82322\ldots, -0.85995\ldots$

Jessie says that the root to the equation $\sin(x + 5) - x = 0$ is $-0.85$ to 2 decimal places. Show that Jessie is correct.

Test the interval [−0.855, −0.845] to show the root lies within it.

$f(-0.855) = \sin(4.145) - (-0.855) = 0.01169\ldots$

$f(-0.845) = \sin(4.155) - (-0.845) = -0.0036391\ldots$

Jessie is correct as there is a change of sign to a continuous function.

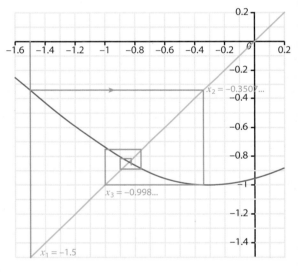

An iterative formula may also diverge. The gradient of its curve can be used to determine if the formula will converge on a specific root or not.

If $-1 < g'(x) < 1$ for all x-values within the given interval, then the formula will converge.

## Example

When approximating the solution to $f(x) = 0$, where $f(x) = x^5 + x - 3$, the root $\alpha$ lies between 1.1 and 1.3. An iterative formula could be found:
$x_{n+1} = 3 - (x_n)^5$

| | |
|---|---|
| $x_1$ | 1.1 |
| $x_2$ | 1.38949 |
| $x_3$ | −2.17937... |
| $x_4$ | 52.1651... |
| $x_5$ | −3.86278982... × $10^8$ |

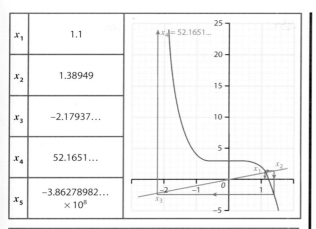

| | |
|---|---|
| $x_1$ | 1.3 |
| $x_2$ | −0.71293 |
| $x_3$ | 3.1841766 |
| $x_4$ | −324.3299... |
| $x_5$ | 3.58868... ×$10^{12}$ |

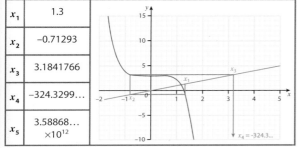

For both starting points this has created a diverging sequence which diverges rapidly. By considering the gradient within the interval:

$$g(x) = 3 - x^5$$
$$g'(x) = -5x^4$$
$$g'(1.1) = -7.3205$$
$$g'(1.3) = -14.2805$$

This shows that the formula $x_{n+1} = 3 - (x_n)^5$ will not converge.

An alternative iterative formula could be tried – this time by rearranging f(x) to make the $x$ from the $x^5$ term the subject of the formula:

$$x^5 + x - 3 = 0$$
$$x^5 = 3 - x$$
$$x = (3 - x)^{\frac{1}{5}}$$
$$x_{n+1} = (3 - x_n)^{\frac{1}{5}}$$
$$g(x) = (3 - x)^{\frac{1}{5}}$$
$$g'(x) = -\frac{1}{5}(3 - x)^{-\frac{4}{5}}$$
$$g'(1.1) = -0.11968...$$
$$g'(1.3) = -0.13081...$$

Both fall within the range −1 < f'(x) < 1, so the formula will give a converging sequence to the root within the interval [1.1,1.3].

| 1.1 | $x_1$ | 1.3 |
|---|---|---|
| 1.13697... | $x_2$ | 1.111961... |
| 1.132514... | $x_3$ | 1.135539... |
| 1.133056... | $x_4$ | 1.132688... |
| 1.132990... | $x_5$ | 1.133035... |
| 1.132998... | $x_6$ | 1.132993... |
| 1.132997... | $x_7$ | 1.132998... |

f(1.13295) = −0.0004394... and
f(1.13305) = 0.00048449...

Change of sign of continuous function so $\alpha$ lies within this interval.

$\alpha$ = 1.1330 (4 d.p.)

> **Don't Forget**
> Layout and communication are important in this topic as small mistakes can make big differences. Make sure the steps taken are clearly expressed and reasoning is fully given.

## Newton-Raphson

The Newton-Raphson method for root-finding uses the gradient of the function to improve the efficiency with which the root is found to any given degree of accuracy. It again uses an iterative approach, producing a sequence of approximations for the root ($\alpha$) to the equation f(x) = 0, but this time the iterative formula is: $x_{n+1} = x_n - \dfrac{f(x_n)}{f'(x_n)}$

The method works where there is a starting point ($x_1$) suitably close to the root and where the function doesn't have asymptotes or stationary points very close to the approximations. Using the ANS button on a calculator makes the process much simpler (see page 43).

### Example

The function $y = f(x)$, where $f(x) = x^7 - x^5 + 0.1$, has two roots ($\alpha$ and $\beta$, $\alpha < \beta$) in the interval [0.4, 1.2].

**a)** Find an approximation, correct to 3 decimal places, for the root $\beta$ using the Newton-Raphson method.

$$x_{n+1} = x_n - \frac{f(x_n)}{f'(x_n)} = x_n - \frac{(x_n)^7 - (x_n)^5 + 0.1}{7(x_n)^6 - 5(x_n)^4}$$

$x_1 = 1.2$ (as this is going to be closer to $\beta$ than $\alpha$)

$x_2 = 1.08656\ldots$

$x_3 = 1.004469\ldots$

$x_4 = 0.952483\ldots$

$x_5 = 0.927952\ldots$

$\left.\begin{array}{l} x_6 = 0.9221453\ldots \\ x_7 = 0.9218339\ldots \end{array}\right\}$ Both give 0.922 to 3 d.p. Test the interval:

$f(0.9215) = -0.000227\ldots$ and
$f(0.9225) = 0.0004594\ldots$
Change of sign of continuous function so a root is 0.922 (3 d.p.)

**b)** The sketch shows $y = f(x)$, with the orange lines representing the first three approximations using $x_1 = 0.4$ with the Newton-Raphson method. The gradient at 0.4 is such that the sequence will converge on $\beta$ again rather than on $\alpha$.

By considering the location of the turning point, suggest a better value for $x_1$ in order to approximate $\alpha$ to 3 decimal places.

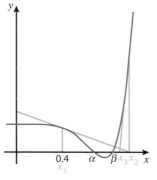

$f'(x) = 7x^6 - 5x^4$

Stationary point when $7x^6 - 5x^4 = 0$ and looking for one within the interval [0.4, 1.2] (so ignore $x = 0$ and $x < 0$ possibilities)

$7x^6 = 5x^4$

$x^2 = \frac{5}{7}$

$x = +\sqrt{\frac{5}{7}} = 0.84515\ldots$

A sensible $x_1$ needs to be well away from the turning point and the value 0.4, but within the range [0.4, 0.8452].

Consider a point comfortably within the interval, e.g. 0.6.

$$x_{n+1} = x_n - \frac{(x_n)^7 - (x_n)^5 + 0.1}{7(x_n)^6 - 5(x_n)^4}$$

$x_1 = 0.6$

$x_2 = 0.756292\ldots$

$x_3 = 0.738174\ldots$

$\left.\begin{array}{l} x_4 = 0.738892\ldots \\ x_5 = 0.738893\ldots \end{array}\right\}$ Both give 0.739 (3 d.p.). Test the interval:

$f(0.7385) = 0.0001383\ldots$ and
$f(0.7395) = -0.0002127\ldots$
Change of sign of continuous function so $\alpha = 0.739$ (3 d.p.)

If the chosen value still didn't give a converging sequence, it is worth using that information to help support another choice for $x_1$. (In this case any value between ~0.524 and ~0.832 would converge to the root $\alpha$: 0.6, 0.7 and 0.8 being the values to 1 d.p. that would converge.)

## Trapezium Rule for Estimating Integrals

You can estimate the area beneath a graph using the trapezium rule. This is important because it isn't always possible to integrate a given function.

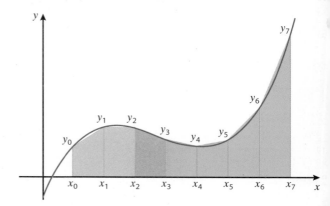

By splitting the shape using a set of ordinates (the vertical lines parallel to the $y$-axis and meeting the $x$-axis, assuming the approximation is an integral with respect to $x$), you can draw in a set of trapezia.

If there are $p$ ordinates, there will be $(p-1)$ strips forming $(p-1)$ trapezia.

If the graph is convex between the ordinates, the trapezium will give an overestimate for that section (shown in red).

If the graph is concave between the ordinates, the trapezium will give an underestimate for that section (shown in green).

The middle section (blue), which is between $x_2$ and $x_3$, contains the point of inflection; this means it could be either an overestimate or an underestimate.

To find the area of an individual trapezium: $\frac{1}{2}(a+b)h$

The trapezium rule considers the sum of all the trapezia formed by using the ordinates and the corresponding $y$-values.

$$\text{Area} \approx \frac{1}{2}h\left\{(y_0 + y_n) + 2(y_1 + y_2 + \ldots + y_{n-1})\right\}$$

To find $h$ when approximating $\int_a^b y\,dx$ with $p$ ordinates, use $h = \dfrac{b-a}{p-1}$

A table of values can be a good way to organise the working and the calculator memory can be very useful for recording accurate values for the different ordinates and saving time.

---

**Example**

Estimate the integral $\int_{0.4}^{1.4} 3x^3 + \cos^2(4x)\,dx$ using the trapezium method with six ordinates and giving your answer accurately to 4 significant figures.

$h = \dfrac{1.4 - 0.4}{6-1} = 0.2$

| $x_0$ | $x_1$ | $x_2$ | $x_3$ | $x_4$ | $x_5$ |
|---|---|---|---|---|---|
| 0.4 | 0.6 | 0.8 | 1.0 | 1.2 | 1.4 |
| 192852... | 1.191749... | 2.532592... | 3.427249... | 5.191656... | 8.833502... |
| (A) in calc. memory | (B) | (C) | (D) | (E) | (F) |
| $y_0$ | $y_1$ | $y_2$ | $y_3$ | $y_4$ | $y_5$ |

$$\int_{0.4}^{1.4} 3x^3 + \cos^2(4x)\,dx$$

$$\approx \frac{1}{2} \times 0.2\{(A+F) + 2(B+C+D+E)\}$$

$$\approx 0.1\{33.71285106\ldots\}$$

$$\approx 3.371285106\ldots$$

$$\approx 3.371 \text{ (4 s.f.)}$$

**b)** Suggest how a better estimation could be obtained for the integral.

By increasing the number of ordinates used, the approximation becomes more accurate.

**c)** Using double angle formulae, show that $3x^3 + \cos^2(4x) = 3x^3 + a + b\cos 8x$ and hence find the percentage error in the estimation from the rounded answer to part **a)**.

$$3x^3 + \cos^2(4x) = 3x^3 + \frac{1}{2}(\cos 8x + 1)$$

$$= 3x^3 + \frac{1}{2} + \frac{1}{2}\cos 8x$$

$$\int_{0.4}^{1.4} 3x^3 + \frac{1}{2} + \frac{1}{2}\cos 8x\,dx$$

$$= \left[\frac{3}{4}x^4 + \frac{x}{2} + \frac{1}{16}\sin 8x\right]_{0.4}^{1.4}$$

$$= \left(\frac{3}{4}(1.4)^4 + \frac{(1.4)}{2} + \frac{1}{16}\sin 8(1.4)\right)$$

$$- \left(\frac{3}{4}(0.4)^4 + \frac{(0.4)}{2} + \frac{1}{16}\sin 8(0.4)\right)$$

$$= 3.5200\ldots - 0.2155\ldots = 3.30445 \text{ (6 s.f.)}$$

Percentage error $= \dfrac{3.30445\ldots - 3.3712\ldots}{3.30445\ldots} \times 100$

$$= -2.02\% \text{ (3 s.f.)}$$

**d)** Ella increases the number of ordinates and sketches the graph, as shown. Suggest why the increase in ordinates may not give a more accurate approximation.

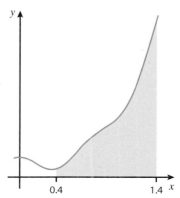

Because the area being found includes both convex and concave sections, there will be some cancelling out of the error in the trapezia. For this reason, the increase in the number of ordinates won't necessarily correlate directly with an improved estimation as it would if the whole interval was either concave or convex.

**Links to Other Concepts**
- Algebraic manipulation
- Graph sketching ● Polynomials
- Indices
- Differentiation and integration
- Problem solving ● Rates of change
- Trigonometry

## Context-based Questions

A question may expect you to select the most appropriate method of approximation. Look out for words like 'approximation' and 'estimate' to suggest that it is a numerical methods question.

## SUMMARY

- **A change of sign of a function indicates an odd number of roots within the interval, assuming the function is continuous within the interval being considered.**

- **No change of sign of the function could indicate no roots, or an even number of roots (repeat roots counting), within the interval.**

- **An iterative equation is a rearrangement of the equation with $x$ as the subject and in terms of $x$. This produces a formula in the form $x_{n+1} = f(x_n)$.**

- **Inputting a value from a known interval may lead to a sequence of approximations that:**
  - **is convergent: as $n \to \infty$, $x_n \to \alpha$, where $\alpha$ is the root to the equation**
  - **is divergent: as $n$ increases, the subsequent terms of the sequence get further away from the root**
  - **may converge to a different root than expected; in this case a different iterative formula can often be found.**

- **To draw a staircase or cobweb diagram, plot the graphs of $y = x$ and $y = f(x)$, where $x_{n+1} = f(x_n)$ is the iterative formula being applied. Starting at the value for $x_1$, draw a vertical line to connect with $f(x)$ and join this horizontally to meet the diagonal line at $x_2$. Repeat this process.**

- **To find an answer to a given accuracy using the iterative method, continue until $x_{n-1}$ and $x_n$ would round to the same value at the given accuracy. Then check that a root does lie within the interval.**

- **Newton-Raphson method: $x_{n+1} = x_n - \dfrac{f(x_n)}{f'(x_n)}$**
  - **Differentiate the function.**
  - **Select an appropriate first approximation (this may be given as part of the question).**
  - **Apply the iteration the required number of times (or to the required accuracy).**
  - **Check the root lies within the interval created by the numbers.**

- **Trapezium rule (to approximate an integral):**
  - **Split the area into a set of ordinates (vertical sections of graph).**
  - **Area $\approx \dfrac{1}{2} h\{(y_0 + y_n) + 2(y_1 + y_2 + \ldots + y_{n-1})\}$**
  - **If a question asks for $p$ ordinates, it will be $(p-1)$ strips.**
  - **To find $h$ when approximating $\displaystyle\int_a^b y \, dx$ with $p$ ordinates, use $h = \dfrac{b-a}{p-1}$**

## QUICK TEST

1. Show that a root to each equation lies within the given interval. Justify your answers.

   a) $3x^2 + 2\cos x - 3 = 0$     [0.5, 0.9]

   b) $\sin(4x) + \cos^2(3x^2) - 1$     [0.5, 0.6]

   c) $\dfrac{6x^2 - 2}{(x+2)(x+4)}$     [−0.6, −0.5]

2. Rearrange each of the following to give two different iterative formulae for finding an approximation to a root.

   a) $x^3 + 3x - 1 = 0$

   b) $x^6 + \sin 4x = 0$

3. Use the Newton-Raphson method to give the first three root approximations for each of these equations:

   a) $4x^2 + \ln x = 0$, $x_1 = 2$

   b) $\frac{1}{2}\cos(4x) + x^3 = 0$, $x_1 = -0.8$

   c) $x^3 \ln x - 3 = 0$, $x_1 = 4$

4. James says the following are approximations to the stated accuracy. Is he definitely correct or incorrect, or is it impossible to tell? Justify your answers.

   a) $4x^2 + \ln x = 0$ has a root which is 0.4480 (4 d.p.)

   b) $\frac{1}{2}\cos(4x) + x^3 = 0$ has a root which is −0.3680 (4 s.f.)

   c) $x^3 \ln x - 3 = 0$ has a root which is 4.917 (4 d.p.)

5. Sam uses the trapezium rule to approximate the integral $\int_{2}^{8} \ln(x^4)\,dx$.

   a) Will his estimate be an overestimate or an underestimate?

   b) If he uses five ordinates, what will the value of $h$ be?

   c) Express the value of $y_1$ and $y_n$ in the form $p \ln 2$.

## PRACTICE QUESTIONS

1. a) $\cos(x^2) - x^{\frac{3}{4}} = 0$. Show that a root to the equation lies in the interval $\left[0, \dfrac{\pi}{3}\right]$.   **[3 marks]**

   b) Find the value of $p$ in the iteration formula $x_{n+1} = (\cos(x_n)^2)^p$.   **[2 marks]**

   The graph shows the lines $y = x$ and $y = (\cos(x^2))^p$.

   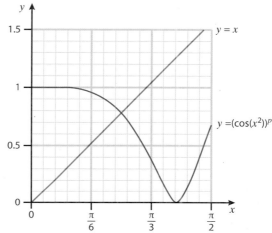

   c) Using $x_1 = 0$, draw a staircase or cobweb diagram to demonstrate the first four approximations of the iterative process.   **[3 marks]**

   d) Find the terms in the sequence up to $x_6$.   **[3 marks]**

   e) Select an appropriate method to estimate the root accurately to 4 decimal places.   **[8 marks]**

2. a) Use the trapezium rule to approximate the integral $\int_{30}^{60} \tan(x^{\frac{1}{10}}) + 3\,dx$ using five ordinates.

   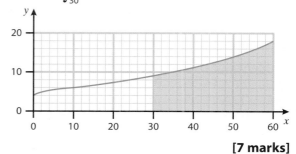

     **[7 marks]**

   b) Describe how you would find a better approximate using the trapezium rule, and whether you would expect your answer to increase or decrease.   **[2 marks]**

# Modelling, Quantities and Units

## The Fundamental Units

**Length** measures a one-dimensional distance between two points. The SI base unit for length is **metres** (m). Distance is a scalar quantity. Displacement is a vector quantity and, as such, has a direction associated with it both when it is positive and negative.

You should know how to convert kilometres to metres (and vice-versa); centimetres to metres (and vice-versa) and millimetres to metres (and vice-versa).

**Time** is measured using the SI base unit **seconds** (s). You should know how to convert minutes to seconds (and vice-versa) and hours to seconds (and vice-versa).

**Mass** is the measure of how much matter there is in an object. It is often confused with weight. An object's mass is constant; it would have the same mass on the Earth or on the Moon. It is measured in SI base unit **kilograms** (kg).

You should know how to convert grams to kilograms (and vice-versa) and tonnes to kilograms (and vice-versa).

## Derived Quantities

All other units come from these basic measurements. The key derived units in mechanics are:

⬤ **Velocity:** measured in **metres per second** ($ms^{-1}$). Velocity is the vector value of speed (i.e. speed but with a set direction). In SI units it is the number of metres travelled every second. It is the rate of change of displacement with regard to time.

⬤ **Acceleration:** measured in **metres per second squared** ($ms^{-2}$). Acceleration measures the rate of change of velocity with respect to time.

⬤ **Force/weight:** measured in **Newtons** (N). Weight is the force an object exerts vertically downwards due to gravity. It is mass multiplied by the acceleration due to gravity. Newtons are equivalent to $kg\,ms^{-2}$, but as force is used so often it is given its own unit of measurement, the Newton. Newtons are used to measure all forces, not just those due to gravity. It is the unit force needed to accelerate a mass of 1 kg at a rate of 1 metre per second squared.

⬤ **Moments:** measured in Newton-metres (Nm). Moments measure the turning force(s) applied about a given point. It is the perpendicular distance between any given point and the force.

## Converting Derived Quantities

By considering the separate parts of the derived quantity, it is possible to convert the units.

> **Example**
>
> Convert $14.4\,kmh^{-1}$ into $ms^{-1}$.
>
> $$\frac{km}{h} \xrightarrow{\times 1000} \times \frac{1000}{3600} \to \frac{m}{s}$$
>
> $$\times 3600$$
>
> $$14.4\,kmh^{-1} = 14.4 \times \frac{1000}{3600} = 4\,ms^{-1}$$

## Dimensional Analysis

When using equations, they need to make sense in terms of the quantities for each side – this can be a useful way of checking if a formula has been remembered correctly.

> **Example**
>
> Rizwan tries to write down the equations of linear acceleration from memory as part of his revision:
>
> $v^2 = u^2 + 2a$    [Equation 1]
>
> $s = ut + \frac{1}{2}at^2$    [Equation 2]
>
> $v = ut + at$    [Equation 3]
>
> By considering the units, show that at least two of these equations have been remembered incorrectly.
>
> Equation 1:
>
> ⬤ LHS: $v^2$ is velocity-squared $\therefore$ measured in $(ms^{-1})^2 = m^2 s^{-2}$.
>
> ⬤ RHS: first term $u^2$ is velocity-squared $\therefore$ measured in $(ms^{-1})^2 = m^2 s^{-2}$.
>
>   This is consistent with the LHS.
>
> ⬤ RHS: second term $2a$ is an acceleration $\therefore$ measured in $ms^{-2}$.

This is **not** consistent with the rest of the equation, so it cannot be correct. To make it correct, it would need to be multiplied by a length, as ms$^{-2}$ × m = m$^2$s$^{-2}$. It would then be consistent throughout the terms of the equation.

Equation 2:

● LHS: $s$ is a displacement ∴ measured in m.

● RHS: first term $ut$ is a velocity multiplied by time, ms$^{-1}$ × s = m.

This is consistent with the LHS.

● RHS: second term $\frac{1}{2}at^2$ is an acceleration multiplied by time-squared, ms$^{-2}$ × s$^2$ = m.

This is consistent with the previous two terms.

● As all the terms have a unit of metres, the equation is balanced dimensionally.

Equation 3:

● LHS: $v$ is a velocity so measured in ms$^{-1}$.

● RHS: first term $ut$ is a velocity multiplied by time, ms$^{-1}$ × s = m.

This is **not** consistent with the LHS, so the formula cannot be correct.

# The Language of Mechanics

Simplifications and assumptions are made in the modelling of real-life situations. You need to know what the key terminology means. You also need to be able to analyse the effect this might have on the theoretical result from the calculation compared to real-life results.

● Body – is used to describe any object with a mass that is being considered.

● Particle – is a modelling assumption meaning that a body has mass but relatively small size (at A-level many things are modelled as particles, e.g. people, cars). The mass is modelled as acting at a single point. It also means there is no air resistance considered during motion.

● Plane – motion and forces are considered at most in two dimensions. The two-dimensional space is referred to as a plane.

● Light – mass is negligible and so is ignored during calculations.

● Inextensible – the string/rope/cable does not stretch, which means a constant force applied at one end is transferred completely through the string and that the bodies will have the same velocity and acceleration.

● String – a relatively thin cord that joins two bodies. Generally a string is modelled as light and inextensible.

● Smooth – the surface and the body experience no frictional force.

● Rough – the body and the plane exert a frictional force on each other.

**Links to Other Concepts**
● Solving quadratics
● Integration and differentiation
● Interpreting graphs and context-based questions
● Solving simultaneous equations
● Vectors ● Algebraic manipulation

## SUMMARY

● **Length is measured in metres where 1000 m = 1 km, 0.01 m = 1 cm and 0.001 m = 1 mm.**

● **Time is measured in seconds where 60 s = 1 minute and 3600 s = 60 minutes = 1 hour.**

● **Mass (not weight!) is measured in kilograms where 0.001 kg = 1 g and 1000 kg = 1 T.**

● **Derived measures:**
   – **Velocity is in metres per second (ms$^{-1}$).**
   – **Acceleration is in metres per second squared (ms$^{-2}$).**
   – **Force/weight is in Newtons (N).**
   – **Moments are in Newton-metres (Nm).**

● **Ensure you are familiar with terms and their meanings for modelling assumptions.**

## QUICK TEST

1. Convert each of the following into the given unit and give an example of something they might be used to measure.

   **a)** 2 days, 4 hours and 32 minutes into seconds

   **b)** 4.5 tonnes into kg

   **c)** 36 Ncm into Nm

   **d)** 70 km/hour into ms$^{-1}$

   **e)** 2000 g cm min$^{-2}$ into Newtons (kg ms$^{-2}$)

# Vectors

A vector is a measure with both **magnitude** (size) and **direction**. Vectors are often used to describe the position of objects; at A-level, this can be in 2D or 3D space.

## Vector Components

You can describe a vector by splitting it into two or three components which act at right angles to each other. This is known as resolving a vector into its components.

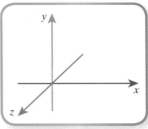

Conventionally, the third axis ($z$) is considered to be coming out of the paper from the $x$ and $y$-axes.

### Writing Vectors in Component Form:
### i, j and k Notation

**i, j** and **k** represent a unit vector in each direction, $x$, $y$ and $z$. A unit vector has length 1. The unit vectors can be multiplied by a constant value, which changes the magnitude but not the direction. However, if the coefficient is negative, the direction is reversed.

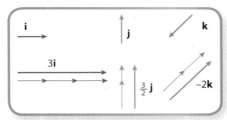

### Example

Describe the vectors **a**, **b** and **c**, as shown on the grid, using **i** and **j** vector notation.

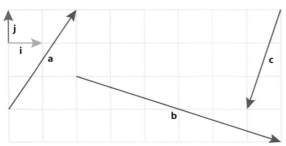

**a:** The line goes two spaces right (+2**i**) and three spaces up (+3**j**).  $a = 2i + 3j$

**b:** The line goes six spaces right (+6**i**) and two spaces down (–2**j**).  $b = 6i - 2j$

**c:** The line goes one space left (–**i**) and three spaces down (–3**j**).  $c = -i - 3j$

## Column Vector Notation

Column notation is a vertical bracket containing two or three numbers $\begin{pmatrix} a \\ b \\ c \end{pmatrix}$ where $a$ represents the 'horizontal' change (positive $x$-direction on a graph), $b$ the 'vertical' change (positive $y$-direction on a graph) and $c$ the 'lateral' change (positive $z$-direction on a 3D set of axes).

Both column notation and **i**, **j** and **k** notation represent the same idea and they can be used interchangeably, unless the question specifies the notation required.

### Example

Describe the vectors **a**, **b** and **c**, as shown on the grid, using column vector notation.

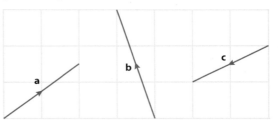

**a:** The line goes two spaces right and one-and-a-half spaces up  $a = \begin{pmatrix} 2 \\ \frac{3}{2} \end{pmatrix}$

**b:** The line goes one space left and three spaces up  $b = \begin{pmatrix} -1 \\ 3 \end{pmatrix}$

**c:** The line goes two spaces left and one down  $c = \begin{pmatrix} -2 \\ -1 \end{pmatrix}$

## Finding the Vector Between Coordinates

To find the displacement vector, $\overline{AB}$, between the points A $(a, b, c)$ and B $(d, e, f)$:

- Change in the $x$-direction goes from $a$ to $d$ and is found using $d - a$.
- Change in the $y$-direction goes from $b$ to $e$ and is found using $e - b$.
- Change in the $z$-direction goes from $c$ to $f$ and is found using $f - c$.

## Example

Find the displacement vector between the points A and B with coordinates (2, 3, –4) and (1, 7, 0) respectively.

Change in $x = 1 - 2 = -1$
Change in $y = 7 - 3 = 4$
Change in $z = 0 - (-4) = 4$

$$\overrightarrow{AB} = \begin{pmatrix} -1 \\ 4 \\ 4 \end{pmatrix} = -\mathbf{i} + 4\mathbf{j} + 4\mathbf{k}$$

## Finding Magnitudes from Components

To find the magnitude of a 2D vector, use Pythagoras' Theorem: $\mathbf{r}^2 = a^2 + b^2 \rightarrow |\mathbf{r}| = \sqrt{a^2 + b^2}$, where $|\mathbf{r}|$ is the magnitude of the vector and $a$ and $b$ are the $\mathbf{i}$ and $\mathbf{j}$ components. Two vertical lines on either side of a vector denote 'the magnitude'.

## Example

$\mathbf{a} = 4\mathbf{i} + 3\mathbf{j}$

$|\mathbf{a}| = \sqrt{4^2 + 3^2} = \sqrt{16 + 9}$

$= \sqrt{25} = 5$

As the value found is scalar (representing size only), the answer is positive.

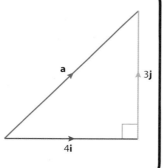

To find the magnitude of a 3D vector, use Pythagoras' Theorem but extended to 3D:
$r^2 = a^2 + b^2 + c^2 \rightarrow |\mathbf{r}| = \sqrt{a^2 + b^2 + c^2}$, where $|\mathbf{r}|$ is the magnitude of the vector and $a$, $b$ and $c$ are the magnitudes of the $\mathbf{i}$, $\mathbf{j}$ and $\mathbf{k}$ components.

## Example

$\mathbf{r} = \begin{pmatrix} 3 \\ -2 \\ 4 \end{pmatrix}$

What is the magnitude of the vector?

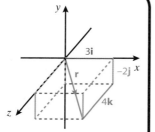

$|\mathbf{r}| = \sqrt{3^2 + (-2)^2 + 4^2} = \sqrt{29}$

## Defining Direction from Components

As shown previously, the components form a right-angled triangle with the vector when considered in two dimensions. To define direction, an angle is needed. So simple trigonometric ratios (normally tan, but cos and sin might be relevant too) are used to define an angle. Ensure you find the required angle.

## Example

The vector $\mathbf{b} = \begin{pmatrix} 2 \\ -5 \end{pmatrix}$. Find the angle between the vector and the unit vector $\begin{pmatrix} 0 \\ 1 \end{pmatrix}$ (often called $\mathbf{j}$).

$\tan \alpha = \frac{\text{opposite}}{\text{adjacent}} = \frac{5}{2}$

$\alpha = \tan^{-1}\left(\frac{5}{2}\right)$

$= 68.1985 \ldots$

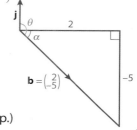

$\theta = \alpha + 90 = 158.2°$ (1 d.p.)

In 3D, angles can be defined for each of the directions $x$, $y$ and $z$. A vector $\mathbf{r} = a\mathbf{i} + b\mathbf{j} + c\mathbf{k}$.

● To find $\alpha$, the angle the vector makes with the $x$-axis, use $\cos \alpha = \dfrac{a}{|\mathbf{r}|}$

● To find $\beta$, the angle the vector makes with the $y$-axis, use $\cos \beta = \dfrac{b}{|\mathbf{r}|}$

● To find $\gamma$, the angle the vector makes with the $z$-axis, use $\cos \gamma = \dfrac{c}{|\mathbf{r}|}$

## Example

$\mathbf{r} = \begin{pmatrix} 3 \\ -2 \\ 4 \end{pmatrix}$. What angle does it make with the $y$-axis?

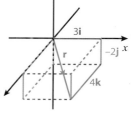

$\cos \beta = \dfrac{-2}{\sqrt{29}}$

$\beta = \cos^{-1}\left(\dfrac{-2}{\sqrt{29}}\right) = 111.8°$ (1 d.p.)

## Resolving Vectors into Components

If given the magnitude and the direction of a vector, you can resolve it into components and express it in either of the forms outlined.

# DAY 6

### Example

A ship sails on a bearing of 030° for a distance of 6 km. Express the displacement as a column vector.

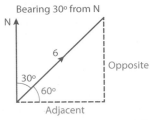

Bearing 30° from N

Horizontal (East)
component $= 6\cos 60$

$= 6 \times \frac{1}{2}$

$= 3$

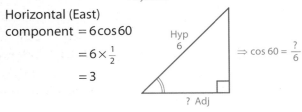

$\Rightarrow \cos 60 = \frac{?}{6}$

Vertical (North)
component $= 6\sin 60$

$= 6 \times \frac{\sqrt{3}}{2}$

$= 3\sqrt{3}$

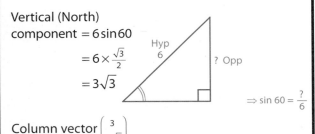

$\Rightarrow \sin 60 = \frac{?}{6}$

Column vector $\begin{pmatrix} 3 \\ 3\sqrt{3} \end{pmatrix}$

## Adding Vectors

A resultant vector (i.e. the overall vector when individual vectors are added together) can be found by placing the individual vectors tail to point. By drawing a diagram, you can see where the right-angled triangles are and use them to find the values of magnitude and direction for the resultant vector.

### Example

Draw the resultant vector of **a** and **b**.

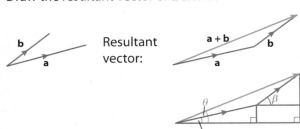

Resultant vector:

A common context is force addition and finding the resultant force on a body.

### Example

A body, modelled as a particle on a smooth, horizontal plane, experiences a force of 3 N at an angle of 30° to the horizontal, and a force of 4 N downwards, as shown in the diagram.

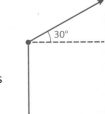

What is the resultant force acting on the body?

Redrawing to represent the vector addition gives:

This shows that the force $F$ has

magnitude $= \sqrt{\left(\frac{3\sqrt{3}}{2}\right)^2 + \left(\frac{5}{2}\right)^2}$

$= \sqrt{13}$

At an angle of $\tan^{-1}\left(\frac{5}{3\sqrt{3}}\right) = 43.89788\ldots$

$= 43.9°$ (1 d.p.) below the horizontal.

$3\sin 30 = \frac{3}{2}$

$3\cos 30 = \frac{3\sqrt{3}}{2}$

$4 - \frac{3}{2} = \frac{5}{2}$

## Multiplication by Scalars

A vector can be multiplied by a constant to create a vector that is effectively enlarged (which can mean getting smaller) by a scale factor.

### Example

$\mathbf{a} = \begin{pmatrix} 5 \\ 0 \\ -2 \end{pmatrix}$

$2\mathbf{a} = 2 \times \begin{pmatrix} 5 \\ 0 \\ -2 \end{pmatrix} = \begin{pmatrix} 2\times 5 \\ 2\times 0 \\ 2\times -2 \end{pmatrix} = \begin{pmatrix} 10 \\ 0 \\ -4 \end{pmatrix}$

## Finding a Unit Vector

To find a unit vector in the direction of the vector

$\mathbf{r} = \begin{pmatrix} a \\ b \\ c \end{pmatrix}$, use $\hat{r} = \frac{r}{|\mathbf{r}|} = \frac{1}{|\mathbf{r}|}\begin{pmatrix} a \\ b \\ c \end{pmatrix}$

## Example

$q = -3i + 12j + 4k$. Find the unit vector in the direction of $q$.

Unit vector $= \dfrac{1}{\sqrt{(-3)^2 + 12^2 + 4^2}}(-3i + 12j + 4k)$

$= \dfrac{1}{13}(-3i + 12j + 4k)$

$= -\dfrac{3}{13}i + \dfrac{12}{13}j + \dfrac{4}{13}k$

## Adding Vectors Algebraically

A defined vector can be added to, and subtracted from, other vectors. Subtraction means reversing the direction suggested by the arrow.

## Example

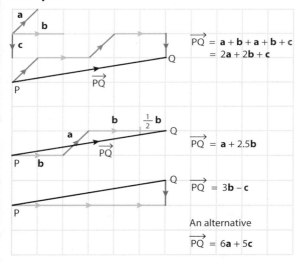

$\overrightarrow{PQ} = a + b + a + b + c$
$= 2a + 2b + c$

$\overrightarrow{PQ} = a + 2.5b$

$\overrightarrow{PQ} = 3b - c$

An alternative
$\overrightarrow{PQ} = 6a + 5c$

## Geometrical Problem Solving

## Example

Regular hexagon ABCDEF has vectors $u = \overrightarrow{AB}$ and $v = \overrightarrow{AD}$.

Using these vectors, state the following:

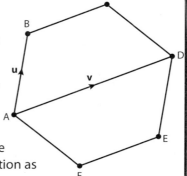

**a)** $\overrightarrow{ED}$

E to D is the same length and direction as A to B, so $\overrightarrow{ED} = u$

**b)** $\overrightarrow{FE}$

F to E is half the length of A to D but is in the same direction, so $\overrightarrow{FE} = \dfrac{1}{2}v$

**c)** $\overrightarrow{AF}$

To go from A to F, consider going to the midpoint of the hexagon then down to F. That is, along $\dfrac{1}{2}\overrightarrow{AD}$ then the equivalent of $\overrightarrow{BA} = -\overrightarrow{AB}$

$\overrightarrow{AF} = -u + \dfrac{1}{2}v$

**Parallel** vectors are in the same direction. To prove that vectors are multiples of each other is to prove that they are parallel.

## Example

$\overrightarrow{AB} = 5i - 3j$ and $\overrightarrow{CD} = 2.5i - 1.5j$. Show that the vectors $\overrightarrow{AB}$ and $\overrightarrow{CD}$ are parallel.

$\overrightarrow{AB} = 5i - 3j = 2(2.5i - 1.5j)$

$\overrightarrow{AB} = 2\overrightarrow{CD}$ so the vectors are parallel.

## Example

The vectors $\begin{pmatrix} 5 \\ -2 \\ \frac{p}{2} \end{pmatrix}$ and $\begin{pmatrix} p \\ -6 \\ q \end{pmatrix}$ are parallel. Find the values of $p$ and $q$.

For the vectors to be parallel $\begin{pmatrix} p \\ -6 \\ q \end{pmatrix} = k\begin{pmatrix} 5 \\ -2 \\ \frac{p}{2} \end{pmatrix}$, where $k$ is a constant.

$\begin{pmatrix} p \\ -6 \\ q \end{pmatrix} = \begin{pmatrix} 5k \\ -2k \\ \frac{p}{2}k \end{pmatrix}$

$-6 = -2k \quad \therefore \quad k = 3$

$p = 5k = 5 \times 3$

$p = 15$

$q = \dfrac{p}{2} \times k = \dfrac{15}{2} \times 3$

$q = 22.5$

**Collinearity** is when three or more points lie on the same straight line. To prove collinearity of points A, B and C, the vectors $\overrightarrow{AB}$ and $\overrightarrow{AC}$ need to be shown to be parallel (i.e. be multiples of each other). If they have a point in common and have the same direction, the three points must lie on a straight line.

# DAY 6

## Example

$\overrightarrow{AB} = -\mathbf{i} - 3\mathbf{j} + \mathbf{k}$ and $\overrightarrow{BC} = -2\mathbf{i} - 6\mathbf{j} + 2\mathbf{k}$

Show that A, B and C are collinear.

$$\overrightarrow{AC} = \overrightarrow{AB} + \overrightarrow{BC}$$
$$= -\mathbf{i} - 3\mathbf{j} + \mathbf{k} + (-2\mathbf{i} - 6\mathbf{j} + 2\mathbf{k})$$
$$= -3\mathbf{i} - 9\mathbf{j} + 3\mathbf{k}$$
$$= 3(-\mathbf{i} - 3\mathbf{j} + \mathbf{k})$$
$$\overrightarrow{AC} = 3 \times \overrightarrow{AB}$$

A, B and C are collinear as they have point B in common.

## Example

The vector diagram shows points A, B, C, D and O.

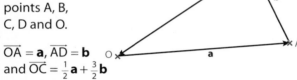

$\overrightarrow{OA} = \mathbf{a}$, $\overrightarrow{AD} = \mathbf{b}$ and $\overrightarrow{OC} = \frac{1}{2}\mathbf{a} + \frac{3}{2}\mathbf{b}$

Given that A, B and C are collinear, find the ratio OB:OD.

$$\overrightarrow{AC} = -\mathbf{a} + \overrightarrow{OC} = -\frac{1}{2}\mathbf{a} + \frac{3}{2}\mathbf{b}$$

As A, B and C are collinear:

$$\overrightarrow{AB} = k\left(-\frac{1}{2}\mathbf{a} + \frac{3}{2}\mathbf{b}\right) = -\frac{k}{2}\mathbf{a} + \frac{3k}{2}\mathbf{b}$$

Let $p$ be fraction $\frac{\overrightarrow{OB}}{\overrightarrow{OD}}$

$$\overrightarrow{OB} = p\overrightarrow{OD}$$

$$\overrightarrow{AB} = \overrightarrow{AO} + p\overrightarrow{OD}$$
$$= -\mathbf{a} + p(\mathbf{a} + \mathbf{b}) = (p - 1)\mathbf{a} + p\mathbf{b}$$

Equate coefficients:

**a:** $p - 1 = -\frac{k}{2}$      **b:** $p = \frac{3k}{2}$

$\quad\quad p = 1 - \frac{k}{2}$

Substitute in variable: $1 - \frac{k}{2} = \frac{3k}{2}$

$1 = \frac{4k}{2} \rightarrow 2k = 1 \rightarrow k = \frac{1}{2}$

$p = 1 - \frac{\frac{1}{2}}{2} \rightarrow p = \frac{3}{4}$

OB:OD $= 3:4$

## Position Vectors

Position vectors relate the position of a point to an origin, much like giving a pair of coordinates on a set of axes.

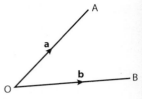

When viewed as a geometrical problem, you can see that to go from A to B along predefined routes would be $-\mathbf{a} + \mathbf{b}$.

To find the position vector of B from A: $\overrightarrow{AB} = \overrightarrow{OB} - \overrightarrow{OA}$

## Example

The points A and B have position vectors $\overrightarrow{OA} = -2\mathbf{i} - 6\mathbf{j}$ and $\overrightarrow{OB} = p\mathbf{i}$.

a) Find, in terms of $p$, the vector $\overrightarrow{BA}$.

$$\overrightarrow{BA} = \overrightarrow{OA} - \overrightarrow{OB} = -2\mathbf{i} - 6\mathbf{j} - p\mathbf{i}$$
$$= (-2 - p)\mathbf{i} - 6\mathbf{j}$$

b) Given that the distance BA = 6.5 km, find the possible values of $p$.

Using Pythagoras' Theorem (other methods are possible):

$$\left(\left|\overrightarrow{BA}\right|\right)^2 = (-2 - p)^2 + (-6)^2$$
$$6.5^2 = 4 + 4p + p^2 + 36$$
$$\frac{169}{4} = p^2 + 4p + 40$$
$$4p^2 + 16p + 160 = 169$$
$$4p^2 + 16p - 9 = 0$$
$$(2p - 1)(2p + 9) = 0$$
$$p = \frac{1}{2} \text{ or } p = -\frac{9}{2}$$

### Links to Other Concepts
- Forces and Newton's laws   • Kinematics
- Motion in a straight line
- Velocity/speed, displacement/distance
- Trigonometry   • Pythagoras' Theorem
- Ratios and fractions   • Algebra
- Geometry   • Graph transformations
- Coordinates   • Simultaneous equations

- **i, j** and **k** are unit vectors in the $x$, $y$ and $z$ directions.

- **Component form of a vector:**

  $\mathbf{p} = a\mathbf{i} + b\mathbf{j} + c\mathbf{k}$ is a vector that represents $a$ units to 'the right' and $b$ units 'upwards' and $c$ units 'outward'.

  p can also be represented by the column vector $\begin{pmatrix} a \\ b \\ c \end{pmatrix}$.

- **To find the magnitude of a vector, |p|**, use **Pythagoras' Theroem:** $|\mathbf{p}| = \sqrt{a^2 + b^2 + c^2}$. Magnitude could be hidden as 'speed', 'distance' or other associated scalar value.

- **To find the angle:**
  - Draw a sketch and identify which angle is being asked for and how to find it.
  - Use **SOH CAH TOA** to calculate the angle.
  - In 3D use $\cos \alpha = \dfrac{a}{|r|}$, $\cos \beta = \dfrac{b}{|r|}$ or $\cos \gamma = \dfrac{c}{|r|}$

- **To find the components:**
  - Draw a diagram and label the given angle.
  - Use **SOH CAH TOA** to find the component vectors.

- **Vectors can be added together:**
  - Using vector algebra
  - If in component form, add the components separately, i.e. $\begin{pmatrix} a \\ b \\ c \end{pmatrix} + \begin{pmatrix} d \\ e \\ f \end{pmatrix} = \begin{pmatrix} a+d \\ b+e \\ c+f \end{pmatrix}$
  - If multiplying by a constant, the magnitude of the vector will be changed but not the orientation. A negative scale factor would reverse the vector's direction.

  $k \times \begin{pmatrix} a \\ b \\ c \end{pmatrix} = \begin{pmatrix} ka \\ kb \\ kc \end{pmatrix}$

- **Position vectors** relate points in space, often to a fixed origin O. To find a relative position, use $\overrightarrow{AB} = \overrightarrow{OB} - \overrightarrow{OA}$.

- **Parallel vectors** are a multiple of each other.

- For points to be **collinear**, the vectors that join one point to each of the others must be multiples of each other.

1. The vector $\mathbf{p} = \begin{pmatrix} -2 \\ 3 \\ 4 \\ \frac{1}{2}\sqrt{2} \end{pmatrix}$

   a) Convert **p** into **i, j, k** notation.

   b) Sketch the vector **p**.

   c) Find the magnitude of the vector **p**.

   d) What angle does **p** make to the unit vector **i**?

2. The displacement vector **p** represents the movement from point A (–3, 12, 0) to point B (5, 7, 3). Write **p** as a column vector.

3. The points A and B have position vectors $\overrightarrow{OA} = 12\mathbf{i} - 7\mathbf{j} + 2\mathbf{k}$ and $\overrightarrow{OB} = -3\mathbf{i} + 7\mathbf{j}$.

   a) Find the vector $\overrightarrow{BA}$.

   b) State the vector $\overrightarrow{AB}$.

## PRACTICE QUESTIONS

1. A mini-submarine travels along a vector $\overrightarrow{AB} = 2a\mathbf{i} + (a - 110)\mathbf{k}$ m, where $a$ is a positive integer greater than 20, followed by a vector $\overrightarrow{BC} = 120\mathbf{i} + 480\mathbf{j} - 160\mathbf{k}$ m. Along these two vectors it travels a total distance of 620 m. **i** is a unit vector to the East, **j** to the North and **k** vertically upwards.

   a) Find the overall displacement vector $\overrightarrow{AC}$. **[7 marks]**

   b) What is the exact distance between points A and C? **[2 marks]**

   The submarine then travels directly to the surface at point D, which is collinear with A and C, and the ratio of the distance AC : AD = 20 : 1.

   c) What is the position vector $\overrightarrow{AD}$? **[5 marks]**

   d) What was the maximum depth of the submarine? **[2 marks]**

# Kinematics

## Graphs

In the exam you may need to use, interpret or produce a graph. If sketching a graph, show all key points clearly. If drawing a graph, plot the points accurately. Remember that time is always on the $x$-axis.

The gradient of a **displacement–time graph** is the rate of change of the displacement with respect to time, which is the **velocity**. Displacement can be positive or negative. To find distance travelled, take the sum of the changes in displacement.

The gradient of a **velocity–time graph** is the rate of change of the velocity with respect to time, which is the **acceleration**. Velocity can be positive or negative.

The area under a velocity–time graph is calculated by multiplying a velocity by time ($\text{ms}^{-1} \times \text{s} = \text{m}$), which gives displacement. If the area is above the $x$-axis, it is positive displacement; if it is under the $x$-axis, it is negative displacement. Distance travelled is the total area between the graph and the $x$-axis taken as a scalar (i.e. all positive) and then added together.

### Sketching a Displacement–Time Graph

● If given a velocity, it is the gradient of the graph. Stationary means zero velocity so the line segment would be horizontal.

● Use given points (e.g. starting at the origin, at time 5 seconds it is 4 m from the origin). Remember that the graph will always join together and never be a vertical line.

● If representing two bodies' motion on a displacement–time graph, they are at the same place at the same time where the lines intersect.

### Example
A car is travelling due East along a straight, horizontal road with a velocity of $14\,\text{ms}^{-1}$. At time $t = 0$ the car passes point P. After 3 seconds the car stops at a set of traffic lights where it has to wait for 20 seconds. The car then continues with a constant velocity and reaches its destination, point Q, which is 340 m East of P at $t = 43$ seconds.

**a)** Sketch a displacement–time graph for its motion.

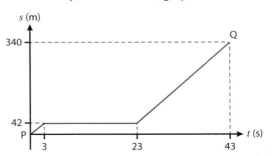

**b)** Find the velocity of the car after the traffic lights.

Gradient = velocity
$$= \frac{340 - 42}{20} = 14.9\,\text{ms}^{-1}$$

**c)** The road has a speed limit of 40 miles per hour. Does the car stay within the speed limit? Use 1609.34 metres = 1 mile.

40 miles per hour $= 40 \times 1609.34$ metres per hour
$$= 64\,373.6 \div 3600$$
$$= 17.881\ldots\text{ms}^{-1}$$

The car stays within the speed limit.

### Sketching a Velocity–Time Graph

● The acceleration is the gradient of the graph.

● If given a displacement or distance travelled, it is the area under the graph.

● If representing two bodies on a velocity–time graph, the intersection between the two lines shows where the bodies have the same velocity at the same time.

● To find where the two bodies meet means equating the areas underneath them (they will have the same displacement at the same time).

### Example
A body is moving along a straight line with a constant velocity of $v = -4\,\text{ms}^{-1}$ for $a$ seconds. It then accelerates with a constant acceleration of $2\,\text{ms}^{-2}$ until it is back at its starting point, at time $t$.

**a)** Sketch a velocity–time graph of its motion.

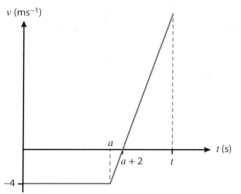

**b)** Find an equation for $t$ in terms of $a$ in the form $t = b\sqrt{a+1} + (a+2)$.

The distance travelled negatively away from the origin is equal to the distance travelled positively back towards the origin. Find the areas of the shapes in terms of $a$ and $t$:

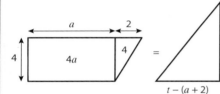

$$4a + 4 = (t-(a+2))^2$$
$$2\sqrt{a+1} = t-(a+2)$$
$$t = 2\sqrt{a+1} + (a+2)$$

**c)** Given that $a = 15$, find the maximum velocity of the body.

Maximum velocity is when time $= t$

$$t = 15 + 2 + 2\sqrt{15+1}$$
$$= 17 + 2\sqrt{16} = 25\,\text{s}$$
$$v = 2(t-(a+2)) = 2(25-(15+2))$$
$$= 16\,\text{ms}^{-1}$$

**d)** A second body starts from rest at the origin and accelerates constantly at $0.5\,\text{ms}^{-2}$. Both bodies continue with their constant acceleration beyond 25 seconds. Find the time when the two bodies are travelling with the same velocity and state the velocity.

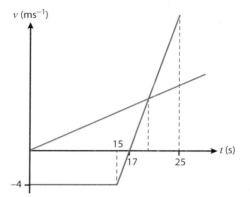

The second body is a straight line through the origin with gradient 0.5, so $v = 0.5t$

The first body will be on the diagonal line, which has gradient 2 and goes through point $(17, 0)$, so $v = 2t - 34$

Equating the two lines: $0.5t = 2t - 34$
$$1.5t = 34$$
$$t = \frac{34}{1.5} = \frac{68}{3} = 22\tfrac{2}{3}\,\text{s}$$

At $22\tfrac{2}{3}\,\text{s}$, both bodies are moving at $11\tfrac{1}{3}\,\text{ms}^{-1}$.

**e)** At what time do the two bodies pass each other?

They pass each other when the displacement from the origin is equal.

Displacement of first body =
$$-64 + \tfrac{1}{2} \times (t-17) \times 2(t-17) = t^2 - 34t + 225$$

Displacement of second body $= \tfrac{1}{2}t(0.5t) = 0.25t^2$

Equal displacement when $0.25t^2 = t^2 - 34t + 225$
$$3t^2 - 136t + 900 = 0$$
$$t = \frac{136 \pm \sqrt{(-136)^2 - 4 \times 3 \times 900}}{2 \times 3}$$
$$t = 37.28\ldots \quad \text{or} \quad t = 8.04\ldots$$
As $t > 17$, $t = 37.3\,\text{s}$   (3 s.f.)

## Equations of Constant Acceleration

When considering motion in a straight line with constant acceleration, you can use the *suvat* equations:

- $s$ = displacement
- $u$ = initial velocity (at the start of the time period)
- $v$ = final velocity (at the end of the time period)
- $a$ = acceleration
- $t$ = time (the duration of the motion)

## Deriving the Equations of Constant Acceleration

You must be able to remember the equations for constant acceleration, and derive them from given assumptions. A classic starting point from which to derive the equations is to use a velocity–time ($v$–$t$) graph.

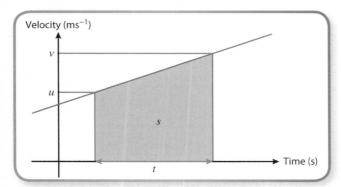

The **area** under a $v$–$t$ graph is the displacement. To find the area, use the area of a trapezium to give:

$$s = \tfrac{1}{2}(u + v)t$$

This formula is used when the acceleration isn't given or asked for.

The **gradient** of the line is the acceleration. To find the gradient, $\frac{\text{change in velocity}}{\text{time for change to occur}}$:

$a = \frac{v - u}{t}$, which rearranges to give $v = u + at$

This formula is used when the displacement isn't given or asked for.

Combining the two previous equations (substituting the second into the first) gives:

$$s = \tfrac{1}{2}(u + u + at)t$$

$$s = ut + \tfrac{1}{2}at^2$$

This formula is used when the final velocity isn't given or asked for.

In a similar way, to create an equation without $u$, rearrange $v = u + at$ to give $u = v - at$, then substitute into $s = \tfrac{1}{2}(u + v)t$ to give:

$$s = \tfrac{1}{2}(v - at + v)t$$

$$s = vt - \tfrac{1}{2}at^2$$

This formula is used when the initial velocity isn't given or asked for.

So the only variable left to be eliminated from a formula is time. By rearranging $v = u + at$ to give $t = \frac{v - u}{a}$, then substituting into $s = \tfrac{1}{2}(u + v)t$, gives:

$$s = \tfrac{1}{2}(u + v)\left(\frac{v - u}{a}\right)$$

$$2as = (u + v)(v - u)$$

$$2as = v^2 - u^2$$

$$v^2 = u^2 + 2as$$

This formula is used when the time isn't given or asked for.

## Using the Equations of Constant Acceleration

The equations can be used to find any one of the variables ($s$, $u$, $v$, $a$ or $t$). The skill is in knowing/deriving the best equation(s), interpreting the context correctly and identifying all the known values accurately (including negatives/positives).

> ### Example
> A helicopter is flying over the sea at a constant height in a straight line. It is accelerating at a constant $0.2\,\text{ms}^{-2}$ towards a lighthouse. At $t = 0$ it is travelling at a speed of $1\,\text{ms}^{-1}$ away from the lighthouse. The lighthouse is 4 km away from the helicopter's starting position. Find the velocity of the helicopter when it is vertically above the lighthouse.
>
> Start by identifying each unknown and whether it is a given, needs to be found or is irrelevant. Check the consistency of units at this point.
>
> Define the positive and negative direction:
>
> Positive direction is from the initial position of the helicopter towards the lighthouse. This could be shown on a diagram.
>
> $s = 4000\,\text{m}$, $u = -1\,\text{ms}^{-1}$, $v = ?$, $a = 0.2\,\text{ms}^{-2}$, $t = $ irrel.
>
> Now select and write the correct equation:
>
> $$v^2 = u^2 + 2as$$
> $$v^2 = (-1)^2 + 2 \times 0.2 \times 4000 = 1601$$
> $$v = \sqrt{1601} = 40.0124\ldots$$
> $$= 40.0\,\text{ms}^{-1}\ (3\ \text{s.f.})$$

## Vector Notation

When working in 2D, vector notation (**i** and **j**, or column vectors) can be an efficient way to show working.

**r** is the **position vector**

$\mathbf{r}_0$ is the position vector at time 0, or the **initial position**

**u** is the **initial velocity**

**v** is the **final velocity**

**a** is the **acceleration**

$t$ is time

Note: Time is not a vector quantity but all the other '$suvat$' letters are vectors.

The equations of constant acceleration can be written in vector form as follows:

- $\mathbf{v} = \mathbf{u} + \mathbf{a}t$
- $\mathbf{r} = \frac{1}{2}(\mathbf{u} + \mathbf{v})t + \mathbf{r}_0$
- $\mathbf{r} = \mathbf{u}t + \frac{1}{2}\mathbf{a}t^2 + \mathbf{r}_0$
- $\mathbf{r} = \mathbf{v}t - \frac{1}{2}\mathbf{a}t^2 + \mathbf{r}_0$

The skill lies in identifying the elements provided by the question and using them to select the appropriate equation to answer it.

### Example

A boat is travelling with constant acceleration of $0.1\,\text{ms}^{-2}$ to the East and $0.2\,\text{ms}^{-2}$ to the South. At 12.07 pm, the boat is 3 km directly East of a lighthouse and has a velocity of $5\,\text{ms}^{-2}$ to the West and $3\,\text{ms}^{-2}$ to the North. What was the position of the boat at 12 noon relative to the lighthouse?

Time between 12.00 and 12.07 in seconds
$= 60 \times 7 = 420$

$\mathbf{a} = \begin{pmatrix} 0.1 \\ -0.2 \end{pmatrix}$, $\mathbf{v} = \begin{pmatrix} -5 \\ 3 \end{pmatrix}$, $\mathbf{r} = \begin{pmatrix} 3000 \\ 0 \end{pmatrix}$, $t = 420$

$\mathbf{r} = \mathbf{v}t - \frac{1}{2}\mathbf{a}t^2 + \mathbf{r}_0$

$\begin{pmatrix} 3000 \\ 0 \end{pmatrix} = \begin{pmatrix} -5 \\ 3 \end{pmatrix} 420 - \frac{1}{2} \times 420^2 \times \begin{pmatrix} 0.1 \\ -0.2 \end{pmatrix} + \mathbf{r}_0$

$\mathbf{r}_0 = \begin{pmatrix} 3000 \\ 0 \end{pmatrix} - \begin{pmatrix} -5 \\ 3 \end{pmatrix} 420 + 88\,200 \times \begin{pmatrix} 0.1 \\ -0.2 \end{pmatrix}$

$= \begin{pmatrix} 3000 \\ 0 \end{pmatrix} - \begin{pmatrix} -2100 \\ 1260 \end{pmatrix} + \begin{pmatrix} 8820 \\ -17\,640 \end{pmatrix}$

$= \begin{pmatrix} 3000 + 2100 + 8820 \\ 0 - 1260 - 17\,640 \end{pmatrix}$

$= \begin{pmatrix} 13\,920 \\ -18\,900 \end{pmatrix} \text{m}$

The initial position (at 12 noon) of the boat is 13.92 km East and 18.9 km South of the lighthouse.

## Calculus and Kinematics

$= \frac{d\mathbf{r}}{dt}$ — Velocity is the rate of change of position with respect to time.

$= \int \mathbf{v}\, dt$ — Position is the integral of the velocity with respect to time.

$= \frac{d\mathbf{v}}{dt} = \frac{d^2\mathbf{r}}{dt^2}$ — Acceleration is the rate of change of velocity with respect to time.

$= \int \mathbf{a}\, dt$ — Velocity is the integral of the acceleration with respect to time.

Remember that if trigonometric functions are to be integrated or differentiated then the angles will be in radians.

### Example

A car moving in a straight line has a velocity of $v = -2t^2 + 4t + 6\,(t \geqslant 0)$.

a) For what values of $t$ does the car have positive acceleration?

$a = \frac{dv}{dt} = -4t + 4 \qquad -4t + 4 > 0$

$4t < 4$

$0 \leqslant t < 1$

b) Find the distance travelled between $t = 1$ and $t = 3$, correct to 2 decimal places.

$r = \int_1^3 -2t^2 + 4t + 6 \ dt$

$= \left[ -\frac{2}{3}t^3 + 2t^2 + 6t \right]_1^3$

$= \left( -\frac{2}{3} \times 3^3 + 2 \times 3^2 + 6 \times 3 \right)$
$\quad - \left( -\frac{2}{3} \times 1^3 + 2 \times 1^2 + 6 \times 1 \right)$

$= (18) - \left( \frac{22}{3} \right)$

$= \frac{32}{3} = 10.67 \ (2\text{ d.p.})$

c) When $t = 1$ the car is at the origin. Find an expression for the displacement from the origin of the car at time $t$.

$r = \int v \ dt = -\frac{2}{3}t^3 + 2t^2 + 6t + C$

$0 = -\frac{2}{3} + 2 + 6 + C$

$C = \frac{2}{3} - 2 - 6 = -\frac{22}{3}$

$r = -\frac{2}{3}t^3 + 2t^2 + 6t - \frac{22}{3}$

**Calculus and Kinematics in Two Dimensions**

When dealing with a problem in 2D, you can use calculus independently on the two parts of the motion.

If $\mathbf{r} = (f(t))\mathbf{i} + (g(t))\mathbf{j}$:

- $v = (f'(t))\mathbf{i} + (g'(t))\mathbf{j}$
- $a = (f''(t))\mathbf{i} + (g''(t))\mathbf{j}$

## Example

A body's velocity can be described as
$v = (t^2 e^{0.2t})\mathbf{i} - \sin(2t)\mathbf{j} \text{ ms}^{-1}$.

**a)** What is the velocity when $t = 0$?

$v = (0^4 e^0)\mathbf{i} - \sin(0)\mathbf{j} = 0 \text{ ms}^{-1}$

**b)** What is the magnitude of the acceleration at $t = 5$?

Use the product rule to differentiate $\left(\frac{d}{dt}(t^2 e^{0.2t})\right)$.

Use inspection or the chain rule to differentiate $\left(\frac{d}{dt}(\sin(2t))\right)$.

$a = \left(\frac{d}{dt}(t^2 e^{0.2t})\right)\mathbf{i} - \left(\frac{d}{dt}(\sin(2t))\right)\mathbf{j}$

$\quad = (2t e^{0.2t} + 0.2 t^2 e^{0.2t})\mathbf{i} - 2\cos(2t)\mathbf{j}$

At $t = 5$,

$a = (2 \times 5 \times e^{0.2 \times 5} + 0.2 \times 5^2 \times e^{0.2 \times 5}) - 2\cos(10)\mathbf{j}$

$\quad = (10e + 5e)\mathbf{i} - 2\cos 10\mathbf{j} = (15e)\mathbf{i} - 2\cos 10\mathbf{j}$

Use Pythagoras' Theorem to find the magnitude. Remember to set your calculator in radians mode.

$|a| = \sqrt{(15e)^2 + (2\cos 10)^2} = 40.8 \text{ ms}^{-2}\,(3 \text{ s.f.})$

**c)** What is the displacement at $t = 5$ in the $\mathbf{j}$ direction from its initial position at $t = 0$?

Displacement in $\mathbf{j}$ direction:

$\mathbf{r_j} = \int -\sin(2t)dt = \frac{1}{2}\cos(2t) + C$

At $t = 0$, $\mathbf{r} = 0$

$0 = \frac{1}{2}\cos(0) + C$

$C = -\frac{1}{2}$

At $t = 5$, $\mathbf{r_j} = \frac{1}{2}\cos(10) - \frac{1}{2} = -0.920 \text{ m} \,(3 \text{ s.f.})$

## Acceleration Due to Gravity

A classic example of motion with constant acceleration is that of objects being dropped or projected. Modelling assumptions are key:

● The object is modelled as a **particle** – this means there is no **air resistance** (if air resistance is included, acceleration would vary with speed). Sometimes this is a reasonable model but at other times not.

● Objects considered as particles do not spin and have no size. Consider the context as to whether this is reasonable or not, and the errors incurred by such assumptions.

● When **projectiles** are launched, it is assumed that the height of release and landing is the same, unless, for example, a diver is jumping from a high board into a pool. The assumption is that the launcher has no vertical height above the 'ground'. If asked to assess the impact of this modelling, again consider the size of the simplification compared to the overall distances involved.

When launched, objects have an initial velocity. This velocity can be split into two components: horizontal ($\mathbf{i}$) and vertical ($\mathbf{j}$). Once released, the object experiences no further forces horizontally and so has a constant horizontal velocity. Vertically it experiences acceleration due to gravity, $g$ ($\sim 9.8 \text{ ms}^{-2}$). Be clear about what direction is positive. If upwards is positive, then $g = -9.8 \text{ ms}^{-2}$.

## Example

A skydiver jumps from a plane and whilst in freefall can be modelled as a particle. She jumps up and out with a horizontal velocity of $2 \text{ ms}^{-1}$ and with the upwards component of her velocity being $2 \text{ ms}^{-1}$. Use $g = 9.8 \text{ ms}^{-2}$.

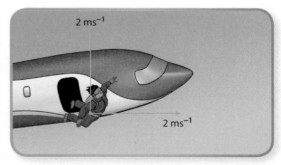

Find the horizontal displacement at the point when the skydiver has a vertical displacement of 378.9 m.

Horizontal displacement $= vt = 2t$
Vertically (positive is downwards):
$s = 378.9$, $u = -2$, $v = $ irrelevant, $a = 9.8$, $t = ?$

$$378.9 = -2t + \frac{1}{2} \times 9.8 \times t^2$$

$$3789 = -20t + 49t^2$$

$$49t^2 - 20t - 3789 = 0$$

$$t = \frac{20 \pm \sqrt{(-20)^2 - 4 \times 49 \times (-3789)}}{98}$$

$t = 9$ ($t \neq -8.591\ldots$ in this context)

Substitute into horizontal:

Displacement $2 \times 9 = 18\,\text{m}$

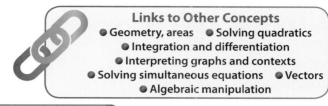

**Links to Other Concepts**
- Geometry, areas • Solving quadratics
- Integration and differentiation
- Interpreting graphs and contexts
- Solving simultaneous equations • Vectors
- Algebraic manipulation

## SUMMARY

- Show key points (intersects, points where the motion changes, etc.) clearly on graphs.
- Displacement–time graphs:
  - Gradient is rate of change of displacement with respect to time, i.e. velocity.
  - Horizontal line means object is stationary.
  - If two lines intersect, the two objects are in the same place at the same time.
  - Curves represent varying velocity; differentiation could be used to find the gradient of the graph at any given point if an equation for the displacement in terms of time is given.
- Velocity–time graphs:
  - Gradient is rate of change of velocity with respect to time, i.e. acceleration.
  - A horizontal line means the object is travelling at a constant velocity.
  - Area between the line and the $x$-axis is the displacement; beneath the $x$-axis it is negative displacement.

- Total area is the total distance travelled.
- If two lines intersect, the two objects have the same velocity at that time.
- To find when two objects are at the same place, equate the displacements and include the negatives if relevant.
- Equations of constant acceleration:

$$s = \tfrac{1}{2}(u+v)t \qquad v = u + at$$
$$s = ut + \tfrac{1}{2}at^2 \qquad r = \tfrac{1}{2}(u+v)t + r_0$$
$$s = vt - \tfrac{1}{2}at^2 \qquad r = ut + \tfrac{1}{2}at^2 + r_0$$
$$v^2 = u^2 + 2as \qquad r = vt - \tfrac{1}{2}at^2 + r_0$$

- Calculus applied to kinematics:

$$v = \frac{dr}{dt} \quad r = \int v\,dt \quad a = \frac{dv}{dt} = \frac{d^2 r}{dt^2} \quad v = \int a\,dt$$

  - In 2D, each element can have calculus applied separately.

- A quadratic, for example when finding $t$ using $s = ut + \tfrac{1}{2}at^2$, will give two answers.

  Acknowledge both, even if one is discounted.

## QUICK TEST

1. Find any unknowns in these:

   a) $r = 2\mathbf{i} + 7\mathbf{j}$, $r_0 = 3\mathbf{j}$, $\mathbf{u} = q\mathbf{i}$, $\mathbf{v} = 2\mathbf{i} - p\mathbf{j}$, $t = 4$

   b) $r = \begin{pmatrix} 50 \\ -20 \end{pmatrix}$, $r_0 = \begin{pmatrix} 34 \\ -52 \end{pmatrix}$, $\mathbf{u} = \begin{pmatrix} 2 \\ 2 \end{pmatrix}$, $\mathbf{a} = \begin{pmatrix} 1 \\ 3 \end{pmatrix}$, $t = ?$

2. $r = (\cos(0.1t^2))\mathbf{i} + (2t\ln t)\mathbf{j}$

   a) What is the expression for the velocity?

   b) What is the expression for the acceleration?

   c) What is the acceleration when $t = 2$?

   d) What is the magnitude of the acceleration when $t = 2$?

## PRACTICE QUESTIONS

1. A body moves on a fixed horizontal plane, where the unit vectors $\mathbf{i}$ and $\mathbf{j}$ represent East and North respectively. The body has a position vector of

   $$r = (3t^2 - 2t)\mathbf{i} + \left( \frac{t}{10}\ln t^2 \right)\mathbf{j}\,\text{m}.$$

   a) What is the acceleration vector in terms of time? **[6 marks]**

   b) What is the speed at $t = 4\,\text{s}$? **[3 marks]**

   c) Find the length of time when the body is within 1 m of the line of longitude (a straight line running North to South), which is 3 m East of the body at $t = 0$. **[5 marks]**

# Mechanics

## Forces

A force is generally a 'push' or a 'pull'. 1 Newton is the force required to accelerate a mass of 1 kg at 1 ms$^{-2}$. The forces met at A-level are:

- **Normal reaction** – a balancing force created by a surface when an object is pressed into it. Importantly, the normal reaction force is always at **right angles to the surface**.

- **Weight** – the force of attraction between two bodies with mass. Weight is the mass of an object multiplied by the acceleration due to gravity, $W = mg$.

- **Tension** – a pulling force created within a 'string' or 'rod'. The pull is experienced equally, but in opposite directions, at each end of the rod. The direction of the force is along the rod.

- **Thrust** – a pushing force. This is a key difference between something modelled as a string (no thrust can be transferred) and a rod (thrust is transferred). The force is transferred through the rod. The thrust from a rod is experienced equally, but in opposite directions, at each end of the rod.

- **Friction** – a force that occurs between a rough surface and an object that are in contact. It acts along the surface in the direction opposite to motion. If an object is held in equilibrium (all forces are balanced), then friction can be a balancing force up to a maximum value of $\mu R$. When the object is moving, friction is equal to $\mu R$.

Force is a vector quantity and, as such, the lines on a force diagram show magnitude and direction. Forces can be represented as any vector, using **i** and **j** notation or column vectors, or by specifying a magnitude and a direction.

## Equilibrium

Forces are in equilibrium when they are balanced, i.e. there is no resultant force. When forces are in equilibrium, there is no acceleration experienced by an object so it is either stationary or continues to move with a constant velocity **(Newton's first law of motion)**.

Using a component vector approach allows two equations to be created showing a balance, or imbalance, of the forces in two directions acting at right angles to each other.

---

**Example**

Four forces, A, B, C and D, are applied to a particle on a smooth horizontal plane.

$$A = \begin{pmatrix} 2 \\ p+1 \end{pmatrix}, B = \begin{pmatrix} p \\ -7 \end{pmatrix}, C = \begin{pmatrix} 2 \\ 0 \end{pmatrix} \text{ and } D = \begin{pmatrix} -6 \\ q \end{pmatrix}$$

Given that the particle is moving with a constant velocity of 3 ms$^{-1}$, find the values of $p$ and $q$.

'$x$-direction': $2 + p + 2 - 6 = 0$
$$-2 + p = 0$$
$$p = 2$$

'$y$-direction': $(p+1) - 7 + 0 + q = 0$
$$3 - 7 + q = 0$$
$$q = 4$$

---

### Normal Reaction, Friction and Equilibrium

Normal reaction is the force that creates equilibrium at right angles to the surface that an object is on.

When resolving forces, it is generally easiest to resolve into components that are parallel and perpendicular to the surface – this is because a number of the forces balance most easily this way. Try to minimise the number of forces that need to be resolved.

With friction, $F_r = \mu R$ only applies when the object is at the point of slipping or in motion. If the body is stationary, then $F_r \leqslant \mu R$. Friction can act either way along a plane but it is opposite to the direction of motion, or potential motion if 'at the point of slipping'.

---

**Example**

A box is sitting on a slope, as shown in the diagram. It is kept in equilibrium, but on the point of slipping down the plane, by a rope that acts along the line of the slope.

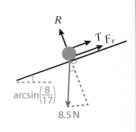

---

**a)** What is the value of the normal reaction force between the plane and the box?

Components perpendicular to the slope are in equilibrium.

Weight component $= 8.5\cos\left(\arcsin\left(\frac{8}{17}\right)\right) = 7.5\,\text{N}$

Resultant force is 0

$R - 7.5 = 0$

The normal reaction $= 7.5\,\text{N}$

**b)** Find the coefficient of friction between the box and the plane given that $T = 1.5\,\text{N}$.

Component of weight $= 8.5\sin\left(\arcsin\left(\frac{8}{17}\right)\right) = 4\,\text{N}$

Resultant force along the plane is 0, so
$F_r + T = 4\,\text{N}$

Point of slipping, so $F_r = \mu R$

$\mu R + 1.5 = 4 \Rightarrow \mu R = 2.5$

$\mu = \frac{2.5}{7.5} = \frac{1}{3}$   (Note: As a coefficient, $\mu$ doesn't have a unit.)

## Resultant Forces

A resultant force is the sum of all the forces acting on a body.

**Example**

The force diagram shows five forces acting on a body. Find the magnitude and the direction, given as a bearing, of the resultant force.

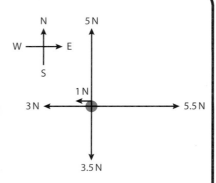

Vertical component of the resultant force
$= 5 - 3.5 = 1.5$ to the North

Horizontal component of the resultant force
$= 5.5 - 3 - 1 = 1.5$ to the East

Overall resultant force:

Magnitude $= \sqrt{1.5^2 + 1.5^2} = \frac{3\sqrt{2}}{2}$

At a bearing of $045°$

## Newton's Second Law of Motion, $F = ma$

Newton's second law states that the resultant force $(F)$ is equal to the mass multiplied by the acceleration of the object. This can be rearranged to find $m$ or $a$:

$$m = \frac{F}{a} \qquad a = \frac{F}{m}$$

**Example**

Louise is being lowered on a winch from a hovering helicopter. She weighs 85 kg, and experiences no external forces other than the tension in the cable. The cable provides a constant tension $T$. Given that Louise accelerates at a rate of $2\,\text{ms}^{-2}$ during the first phase of motion, find the value of the tension in the cable. The winch cable is modelled as light and inextensible. Louise is modelled as a particle. Take $g = 9.8\,\text{ms}^{-2}$.

$85g - T = \text{resultant force}$

$F = ma = 85 \times 2 = 170$

$85 \times 9.8 - T = 170$

$T = 85 \times 9.8 - 170 = 663\,\text{N}$

### Newton's Second Law of Motion with Equations of Constant Acceleration

**Example**

Louise is lowered from the helicopter starting with 0 velocity, and after 4 seconds the tension in the winch cable is increased to 1003 N, which allows her to land on the ground travelling at $4\,\text{ms}^{-1}$. Find the height at which the helicopter is hovering.

For first period of motion:
$s = ?, u = 0, v = ?, a = 2, t = 4$

(Note: Positive direction is being taken as downwards)

To find $v$, use $v = u + at$
$v = 0 + 2 \times 4 = 8\,\text{ms}^{-1}$

To find $s$ use any equation, but if possible use the values given by the question rather than calculated:

$s = ut + \frac{1}{2}at^2$

$s = 0 + \frac{1}{2} \times 2 \times 4^2 = 16\,\text{m}$

For second period of motion:

$F = ma$

$85g - 1003 = 85a$

$a = \frac{85g - 1003}{85} = -2\,\text{ms}^{-2}$

(Note: The negative implies upwards acceleration)

$s = ?, u = 8, v = 4, a = -2, t = $ irrel.

To find $s$, use $v^2 = u^2 + 2as$

$4^2 = 8^2 + 2 \times -2 \times s$

$s = \frac{16 - 64}{-4} = 12\,\text{m}$

Total distance from the hovering helicopter to the ground $= 12 + 16 = 28\,\text{m}$

## Connected Particles

Connected particles can be modelled as two bodies separately. The tension is equal through the string/rod/rope so is felt equally, but possibly in different directions, by each object. Acceleration is also constant assuming the rope stays taut. This often leads to a pair of simultaneous equations to solve in terms of $a$ and $T$.

### Horizontal

A car that pulls a trailer can be modelled as two separate bodies and force diagrams drawn for each.

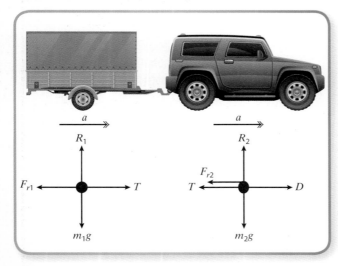

$T$ is the same magnitude in each diagram. The acceleration is also the same for the car and the trailer.

Vertically the vehicle and trailer are in equilibrium – they are travelling along the road. The vertical forces are significant though, as they determine the different frictional forces $F_{r1}$ and $F_{r2}$. It is also possible to model the car and trailer as a single body, as they are travelling together with the same velocity:

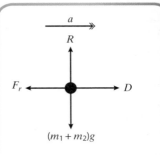

Note: $F_r$ is the combined friction experienced by both vehicles.

### Vertical

Two particles are connected by a string that passes over a smooth pulley (this could be a lift and its counterweight). Again each body can be represented separately. The pulley adds the fact that the acceleration and the motion, whilst still equal in magnitude, are opposite in direction.

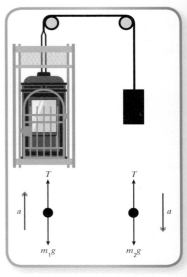

If an extra equation is needed to help to solve the problem then you can consider it as one body, but first the directions need to be aligned so that the acceleration is in the same direction.

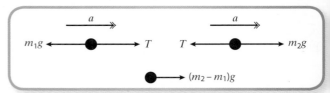

### Don't forget
By modelling the two objects as one body, the tension force is eliminated. Light, inextensible string means the two bodies will have the same acceleration and velocity at all times (as long as the string is taut).

## Example

A child has tied together two toys with a length of string. She places them both on a table, which is 58 cm tall.

The string is taut at the point when the tractor, which has a mass of 270 g, falls off the edge of the table. The teddy, whose mass is 180 g, is pulled along the rough horizontal surface of the table top. Given that $\mu = 0.5$:

**a)** Find the acceleration of the teddy in terms of $g$.

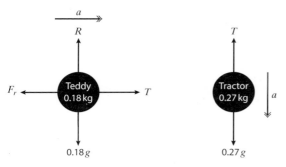

Tractor: $F = ma$

Resultant force $F = 0.27g - T$

$0.27a = 0.27g - T$

Teddy vertically in equilibrium: $R = 0.18g$

Teddy horizontally: $F = ma$

Resultant force $F = T - \mu \times 0.18g = T - 0.09g$

$0.18a = T - 0.09g$

Solve as simultaneous equations (finding both unknowns adds a check in the working):

$T = 0.18a + 0.09g$ (rearrangement of horizontal equation for teddy)

$0.27a = 0.27g - (0.18a + 0.09g)$ (substituted into tractor vertical equation)

$0.27a = 0.18g - 0.18a$

$0.45a = 0.18g$

$a = 0.4g \text{ ms}^{-2}$

**b)** Find the velocity of the tractor as it hits the ground. Use $g = 10 \text{ ms}^{-2}$.

$s = 0.58, u = 0, v = ?, a = 4, t = \text{irrelevant}$

$v^2 = u^2 + 2as$

$v^2 = 0 + 2 \times 4 \times 0.58 = 4.64$

$v = \sqrt{4.64} = 2.15 \text{ ms}^{-1}$ (3 s.f.)

## Connected Particles on a Slope

Questions may bring together different elements: friction, slopes and resolving, acceleration, equations of motion, connected particles, etc.

### Example

Three particles rest on a series of surfaces. Initially, particle A, with mass 6 kg, rests on a smooth plane inclined at 30° to the horizontal. A string joins it to particle B, passing over a smooth pulley. The string is taut. Particle B, with mass 4 kg, rests on a rough horizontal plane. Particle C rests on a rough inclined plane at 45° to the horizontal. It is held such that there is no initial tension in the string that passes over a smooth pulley and attaches to B.

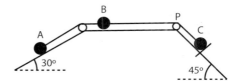

**a)** Particles A and B are in equilibrium but at the point of slipping. Find the value of the coefficient of friction for the horizontal plane.

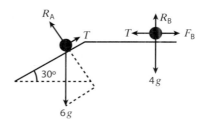

**Particle A (smooth plane: no friction)**

(parallel to plane):  $T = 6g \sin 30$

$= 3g$

**Particle B**

(perpendicular to plane): $\qquad R_B = 4g$

(parallel to plane): $\qquad\qquad F_B = T$

$$\mu \times 4g = 3g$$

$$\mu = \frac{3g}{4g} = 0.75$$

**b)** The support at C is removed. The three particles start to move with a constant velocity of 0.5 ms$^{-1}$. Find the mass of particle C given that the coefficient of friction between C and the inclined plane is 0.4.

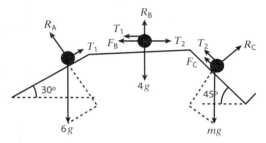

**Particle A**

(parallel to plane): $\qquad T_1 = 6g \sin 30$

$$= 3g$$

**Particle B**

(perpendicular to plane): $\quad R_B = 4g$

(parallel to plane): $\qquad\quad T_2 = T_1 + F_B$

$$= 3g + 4g \times 0.75$$

$$= 6g$$

**Particle C**

(perpendicular to plane): $\quad R_C = mg \cos 45$

$$= \frac{\sqrt{2}}{2} mg$$

(parallel to plane): $\quad mg \sin 45 = T_2 + F_C$

$$= 6g + \frac{\sqrt{2}}{2} mg \times 0.4$$

$$m \times \frac{\sqrt{2}}{2} = 6 + \frac{\sqrt{2}}{5} m$$

$$m\left(\frac{\sqrt{2}}{2} - \frac{\sqrt{2}}{5}\right) = 6$$

$$m = 10\sqrt{2} \text{ kg}$$

$$= 14.1421... \text{ kg}$$

**c)** The string snaps at A when particle B is 0.6 m from the pulley, P. How long does it take B to reach the pulley?

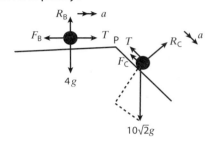

**Particle B**

(parallel to plane): $\qquad 4a = T - 3g$

$$T = 4a + 3g$$

**Particle C**

$$F_B = \mu R_B = 3g$$

(perpendicular to plane): $10\sqrt{2}g \cos 45 = R_C$

$$R_C = 10g$$

(parallel to plane):

$$10\sqrt{2}a = 10\sqrt{2}g \sin 45 - T - 0.4 \times 10g$$

$$10\sqrt{2}a = 6g - T$$

$$= 6g - 4a - 3g$$

$$(10\sqrt{2} + 4)a = 3g$$

$$a = 1.6205... \text{ ms}^{-2}$$

$s = 0.6 \qquad u = 0.5 \qquad s = ut + \frac{1}{2}at^2$

$$0.6 = 0.5t + \frac{1}{2} \times 1.6205... \times t^2$$

$$0.8102...t^2 + 0.5t - 0.6 = 0$$

$$t = \frac{-0.5 \pm \sqrt{0.5^2 - 4 \times 0.8102 \times (-0.6)}}{1.6205}$$

$t = 0.61$ or $-1.22$ (2 d.p.)

Ignore negative time so $t = 0.61$ s

**Links to Other Concepts**
- Solving quadratics
- Trigonometry and geometry
- Integration and differentiation
- Solving simultaneous equations
- Vectors ● Algebraic manipulation
- Interpreting context-based questions

## SUMMARY

- Forces are vectors and can be represented using force diagrams.

- An object is in equilibrium when the resultant force is 0.

- An object in equilibrium will have zero acceleration; this means it will maintain its velocity (which could be stationary).

- If there is a resultant force acting on a body, the body will accelerate in the direction of the force.

- $F = ma$. This second law of motion from Newton is often combined with kinematics questions.

- Friction:

  $F \leqslant \mu R$

  $F = \mu R$ when an object is at the point of slipping, or is in motion

  Friction is in the opposite direction to the motion and along the plane. If an object isn't moving, the friction will be in the opposite direction to the resultant force without the friction.

- Connected particles have the same velocity and acceleration (assuming the connection is taut throughout the motion).

- The tension through the connection is equal in magnitude.

- Bodies that are connected can be modelled as a single body or considered separately.

- Particles on a slope will require some forces to be resolved into their components using trigonometry.

## QUICK TEST

1. Find the resultant force in these diagrams.

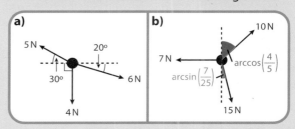

2. For each diagram above, find the acceleration experienced by the object, given that they have the following masses:

   a) 0.2 tonnes    b) $\sqrt{5}$ kg

## PRACTICE QUESTIONS

1. A boy is pulling a sledge (4 kg) along the flat and then up a hill. He pulls the sledge holding the end of a rope which is 73 cm long. Whilst on the horizontal, the height from the boy's hand to the sledge is 55 cm. The boy pulls the sledge with a constant force and along the flat is accelerating at 0.5 ms$^{-2}$.

a) Find an expression for the coefficient of friction between the sledge and the ground in terms of $T$ and $g$.  **[5 marks]**

b) If in this case $0 \leqslant \mu \leqslant 1$, find the possible range of values for $T$. Use $g = 9.8$ ms$^{-2}$.  **[4 marks]**

The tension in the rope is 8 N. Once on the slope, the boy continues to pull with the same force and the coefficient of friction remains constant.

c) How far up the hill will the sledge get before coming to a rest, given that it started the ascent at 8 ms$^{-1}$? Give your answer to 2 significant figures.  **[8 marks]**

# Moments

## Introduction to Moments

Moments deal with turning forces. To find a moment, the force applied is multiplied by the perpendicular distance from the force to the point about which the moment is being considered.

**Example**

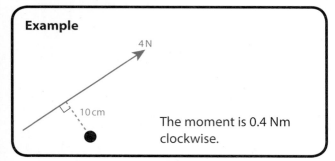

The moment is 0.4 Nm clockwise.

Moments about a point clockwise will cancel out with moments anticlockwise, leaving a resultant (or overall) moment experienced about a given point.

**Example**

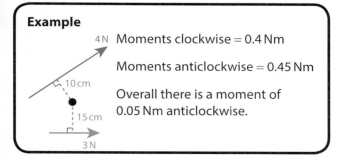

Moments clockwise = 0.4 Nm

Moments anticlockwise = 0.45 Nm

Overall there is a moment of 0.05 Nm anticlockwise.

Moments will generally be used to describe static situations in which there is no overall turning moment.

clockwise moments = anticlockwise moments

$$P \circlearrowright = P \circlearrowleft$$

**Example**

Find the exact value of $P$ given that there is no resultant moment about the point shown.

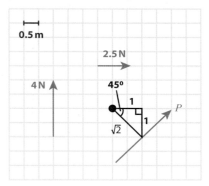

Perpendicular distance to $P$ from the point is found using Pythagoras' Theorem.

Distance is $2 \cos 45 = \sqrt{2}$ m

$$\text{Clockwise} = \text{Anticlockwise}$$

$$P \circlearrowright = P \circlearrowleft$$

$$2.5 \times 1.5 + 4 \times 2 = P \times \sqrt{2}$$

$$P = \frac{47}{4\sqrt{2}} = \frac{47\sqrt{2}}{8}$$

Note: The use of the word 'exact' in the question may be a clue that a surd or other irrational number will be part of the answer. $P$ has been shown as acting in an anticlockwise direction but, if working showed $P$ to be negative, this would be the same as it acting positively in a clockwise direction.

## Resolving Forces

Sometimes it is easier to find the perpendicular distance. In other cases, you may be able to use a given distance but resolve the force into components such that one is perpendicular to the line of distance given.

## Example

To find the moments about the marked point, either use trigonometry to find the perpendicular distance to force or use it to find the perpendicular component of the force. As shown, both have the same overall result of a 5 Nm moment acting anticlockwise about the point:

Perpendicular distance to force:

$$1 \sin 30 = \frac{1}{2}$$

$$\Rightarrow 10 \times \frac{1}{2} = 5\,\text{Nm}$$

Perpendicular component of force from given distance:

$$10 \sin 30 = 5$$

$$\Rightarrow 1 \times 5 = 5\,\text{Nm}$$

The second component of the force (in this example the vertical component) acts directly through the point. As such, it has no perpendicular distance and so no moment is generated by it.

## Context

Questions come in many forms but the most common are about bridges and beams:

- Modelling assumes that the rods/laminas/beams cannot bend and so the turning force is transferred in full.

- Generally the rod/beam/lamina will not be light but, if it is light, then it is treated as having no mass.

- A rod or beam is modelled as one-dimensional and a lamina as two-dimensional.

- Strings tend to be modelled as light and inextensible.

- Freely pivoted means that no moment can be transferred through the point where a rod meets an object. It is possible for the hinged joint to create both vertical and horizontal forces at that point of the rod.

Remember that forces are needed to multiply the distance and that a mass is not a force. $F = mg$

# Centres of Mass of a Body

When dealing with a **uniform** beam or lamina, the assumption is that the mass of the object is spread evenly throughout the object. Therefore, the weight of the object can be considered to act through the centre of mass of the shape, which can be determined by symmetry. If a body is not uniform, then the position of the centre of mass will be given or will be found as part of the question.

The balancing of forces vertically can be used to help find missing pieces of information. This may lead to simultaneous equations to solve.

## Example

A beam is considered to be uniform and sits resting on two supports, as shown. The beam has a mass of 20 kg and length 2 m.

a) By taking moments about support A, find the magnitude of the reaction force exerted at B. (Use $g = 9.8$ ms$^{-2}$)

Adding all the useful forces and distances to the diagram is an important step in the working (shown in blue).

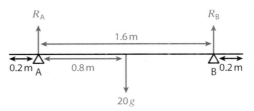

Taking moments about A:

M(A):  $P\circlearrowleft = P\circlearrowright$

$$20g \times 0.8 = R_\text{B} \times 1.6$$

$$R_\text{B} = \frac{156.8}{1.6} = 98\,\text{N}$$

A person stands on the beam at point C and the beam remains in equilibrium. The reaction force at A is now twice that at B.

**b)** Find the distance AC given that the person weighs 64 kg.

Remodelling is very important. Think carefully about what stays the same and what will change when a new load is added (or any other change made). As the person makes the reaction force at A greater than at B, it is reasonable to place it to the left of the centre of the beam:

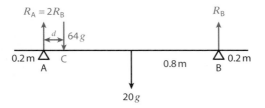

Vertically: $\uparrow = \downarrow$

$$R_A + R_B = 20g + 64g$$

$$3R_B = 20g + 64g$$

$$R_B = \frac{823.2}{3} = 274.4$$

M(A):　　　　$P\circlearrowleft = P\circlearrowright$

$$d \times 64g + 0.8 \times 20g = 1.6R_B$$

$$d \times 627.2 + 156.8 = 1.6 \times 274.4$$

$$d = \frac{439.04 - 156.8}{627.2} = 0.45\,\text{m}$$

AC is 45 cm, with C in between points A and B.

Note: When choosing a point about which to take moments, sometimes one particular point is far more helpful than another. In this case, taking moments about A means that the distance $d$, which is the unknown to be found, will only appear in one term. If moments were taken about C, for example, it would still be solvable but require a little more algebraic manipulation:

M(C):　$d \times 2R_B + (0.8 - d)\,20g = (1.6 - d)R_B$

## Vectors and Moments

Vectors can be used to describe the forces involved in a question.

### Example

The points A, B, and C lie on a coordinate grid: A(3, 4), B(−4, −3), C(1, −p).

The following forces are applied at each point:

$A\begin{pmatrix} 2 \\ -3 \end{pmatrix}$, $B\begin{pmatrix} 3 \\ 1.8 \end{pmatrix}$, $C\begin{pmatrix} p \\ p \end{pmatrix}$, where $p > 0$

The moments taken around point D(1, −2) are in equilibrium. Find the resultant moment about point A.

A sketch is often a good starting point:

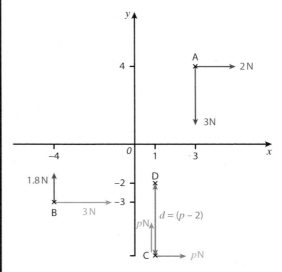

The first thing to find is the value of $p$ using the statement of equilibrium:

M(D):　　　　　　$P\circlearrowleft = P\circlearrowright$

$$2 \times 6 + 3 \times 2 + 1.8 \times 5 = 3 \times 1 + p \times (p - 2)$$

Note: the vertical component of the force, $p$, at C is straight through the pivot point, so produces 0 moment.

$$27 = 3 + p^2 - 2p$$

$$p^2 - 2p - 24 = 0$$

$$(p + 4)(p - 6) = 0$$

As $p > 0$, $p = 6$

$\circlearrowleft$M(A): Resultant moment =
$(1.8 \times 7) - (3 \times 7) + (6 \times 2) - (6 \times 10) = -56.4$ Nm

So the moments about point A are 56.4 Nm anticlockwise.

## Problem Solving

Exam questions may require interpretation, assumptions, reasoning and problem-solving skills. Try to select the appropriate skills that apply. Forces and free body diagrams are important.

### Example
A shop sign measuring 40 cm by 25 cm is suspended by two light, inextensible strings and supported by a pair of rods, as shown in the diagram. The sign has mass $m$ kg and is modelled as a uniform lamina. The sign hangs in equilibrium.

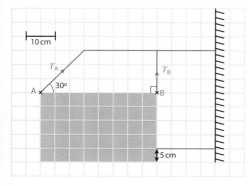

**a)** Select a point about which to take moments that will give an expression for $T_A$ in terms of $T_B$.

Point vertically below centre of mass and horizontally in line with the lower rod. Let the point be point Q.

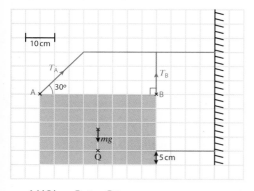

$M(Q): \quad P\circlearrowleft = P\circlearrowright$

$$0.2T_B = 0.2T_A \cos30 + 0.2T_A \sin30$$

$$T_B = \left(\frac{1 + \sqrt{3}}{2}\right)T_A$$

**b)** By considering vertical equilibrium, find an expression for $m$ in terms of $T_A$ and $g$.

Vertically $\uparrow = \downarrow$

$$T_A \sin30 + T_B = mg$$

$$mg = \frac{1}{2}T_A + T_B$$

$$= \left(\frac{1}{2} + \frac{1 + \sqrt{3}}{2}\right)T_A$$

$$= \frac{2 + \sqrt{3}}{2}T_A$$

$$m = \frac{(2 + \sqrt{3})T_A}{2g}$$

**c)** Given that $m = 0.8$ kg, what is the value of $T_B$ to 2 significant figures?

$$T_A = 0.8 \times 2 \times 9.8 \div (2 + \sqrt{3}) = 4.2014433\ldots$$

$$T_B = 4.2014433\ldots \times \left(\frac{1 + \sqrt{3}}{2}\right) = 5.739278\ldots$$

$$= 5.7 \text{ N (2 s.f.)}$$

## Tilting

A beam or rod is about to tilt when there becomes only one point giving a reaction. This could be because:

- it is about to tilt about support A and so support B will have 0 reaction force:

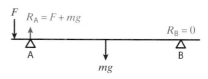

- of a tilt from one suspended point:

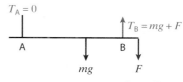

- it is projecting out from a surface and will tilt about the corner of the surface:

## Example

A student places a 30 cm ruler overhanging the edge of a desk. She places her eraser on one end and then gently pushes the ruler out as far as possible. At the point of tilting, 19.5 cm of the ruler is overhanging. The eraser has a mass of 15 g.

**a)** Find the mass of the ruler by modelling it as a uniform rod (with exact length 30 cm).

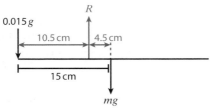

M(point of tilting):   $P\circlearrowleft = P\circlearrowright$

$$0.045 \times mg = 0.105 \times 0.015 \times g$$
$$m = \frac{0.105 \times 0.015}{0.045} = 0.035\,\text{kg}$$
$$m = 35\,\text{g}$$

On weighing the ruler to check her result, the student finds it weighs 25 g. She decides this is because the ruler has a hole cut in one end, meaning it isn't uniform.

**b)** Find the reading at the centre of mass of the ruler.

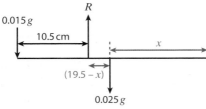

Vertically $\uparrow = \downarrow$

$R = 0.015g + 0.025g = 0.04g$

M(0 cm on ruler):   $P\circlearrowleft = P\circlearrowright$

$0.195 \times 0.04g = x \times 0.025g + 0.3 \times 0.015g$

$$x = \frac{0.195 \times 0.04 - 0.3 \times 0.015}{0.025} = 0.132\,\text{m}$$

The reading on the ruler at the centre of mass is 13.2 cm.

**c)** What assumptions have been made?

The ruler is a rod, i.e. doesn't bend and only has one dimension.

The eraser can be treated as a point mass exactly at the end of the ruler.

The ruler has a length of exactly 30 cm (often 30 cm rulers are slightly longer).

### Links to Other Concepts
- Vectors
- Simultaneous equations
- Forces
- Trigonometry
- Diagrams
- Algebra and quadratics
- Modelling and quantities

## SUMMARY

- A moment is the force multiplied by the perpendicular distance: $P = Fd$

- When in equilibrium, moments and forces are balanced:

  $P\circlearrowleft = P\circlearrowright$    $\uparrow = \downarrow$    $\leftarrow = \rightarrow$

- A beam on the point of tilting will have only one reaction force.

- The centre of mass of a uniform lamina or a beam can be found using the symmetry of the shape.

- Remember to remodel when a situation changes – think carefully about what will stay the same (e.g. mass of the beam) and what will change (e.g. reaction forces).

1. Find the total moments about A in each diagram:

a)
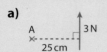

b)
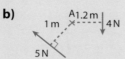

c)

d)
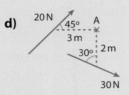

2. The moments in each diagram are in equilibrium. Work out the missing values.

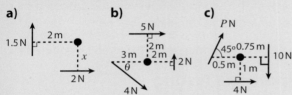

a)   b)   c)

3. The following forces are applied at the stated coordinates. Find the total moments about the origin.

a) Force A $\begin{pmatrix} 3 \\ -5 \end{pmatrix}$ N at (1, 0) and force B $\begin{pmatrix} -2 \\ 1 \end{pmatrix}$ N at (3, −3)

b) Force A $\begin{pmatrix} 10 \\ 3 \end{pmatrix}$ N at (2, −1) and force B $\begin{pmatrix} 4 \\ 2 \end{pmatrix}$ N at (−2, 3)

c) Force A $\begin{pmatrix} -1 \\ 1.5 \end{pmatrix}$ N at (−2, 0), force B $\begin{pmatrix} 4 \\ 5 \end{pmatrix}$ N at (2, 2) and force C $\begin{pmatrix} 1 \\ -3 \end{pmatrix}$ N at (1, −2)

4. Find the missing value(s) given that the beams sit in equilibrium:

a)
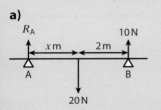

b)
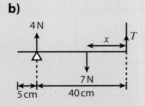

## PRACTICE QUESTIONS

1. A children's playground has a small bridge made by a plank sitting on a support at A and suspended by a rope at B. The plank weighs 500 N.

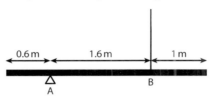

a) By modelling the plank as uniform, find the values of the reaction at A and the tension in the rope at B when the bridge is unloaded.
**[4 marks]**

b) Before being used, the bridge is tested to the limits. What is the greatest single point load the bridge can safely take to prevent tilting?
**[5 marks]**

c) The designer needs the bridge to be safe for a single point mass of up to 150 kg. To achieve this, they move A and B towards the ends of the bridge. What will be the minimum distance AB (rounded to 2 s.f.) for a safe load of 150 kg? Take $g = 9.8$ ms$^{-2}$.
**[5 marks]**

2. A child creates a collage of a train by sticking rectangular pieces of card onto a larger sheet of card and drawing on two circles. The picture is hung on the wall by two pieces of string with pegs at A and B. The card is considered to be a uniform lamina and the strings light and inextensible. The weight of glue and ink are considered to be insignificant.

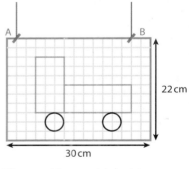

The card is 300 gm$^{-2}$. The mass of each peg is 5 g.

Find the value of the tension in each string.
**[10 marks]**

# Correlation

Building on AS-level work, this topic looks at changing the variable within a regression line and looks deeper into correlation and hypothesis testing for correlation coefficients.

## Regression: Change of Variable

Change of variable is also known as **coding**, which was covered at AS-level for the mean and standard deviation. A regression line could be given in two variables but may be required in terms of a different two variables.

> **Example**
> The regression line of $y$ on $x$ is given as $y = 3 + 5x$. Find the regression line of $p$ on $q$ where $q = 2x$ and $y + 100 = p$.
>
> Rearrange to give $x = \frac{q}{2}$ and $y = p - 100$
>
> Substitute to give $p - 100 = 3 + 5\left(\frac{q}{2}\right)$
>
> Regression line of $p$ on $q$ is $p = 103 + \frac{5q}{2}$

The variables may be linked by exponential equations; in this case, take logs of both sides to change it into a linear regression line. The variables may be part of the exponent.

> **Example**
> The variables $x$ and $y$ are linked by the equation $y = ab^x$. Values of $\log y$ were plotted against $x$ and the regression line was $\log y = 0.1 + 0.2x$. Find $a$ and $b$.
>
> Take logs of both sides for $y = ab^x$ to give $\log y = \log a + x \log b$
>
> Compare coefficients:
>
> $\log y = 0.1 + 0.2x$
> $\quad\quad 0.1 = \log a \rightarrow a = 10^{0.1} \rightarrow \quad a = 1.26$
>
> $\log y = \log a + x \log b$
> $\quad\quad 0.2 = \log b \rightarrow b = 10^{0.2} \rightarrow \quad b = 1.58$

## The Product Moment Correlation Coefficient (PMCC)

At AS-level, correlation was described in terms of positive or negative and in the context of the question. The PMCC, denoted by $r$, is a measure of how strong the correlation is, i.e. how close to a straight line the points lie:

- ⬤ $r = +1$ would be a perfect positive correlation
- ⬤ $r = -1$ would be a perfect negative correlation
- ⬤ $r = 0$ would be no correlation

In reality $r$ will be a decimal value where $-1 \leqslant r \leqslant +1$. The strength of the correlation depends on how close to each of the extreme values $r$ is. It is expected that $r$ is calculated using calculator functions.

The value of the PMCC is not affected by coded data.

## Hypothesis Testing

Hypothesis tests are used to check whether there is evidence that two variables are correlated. The null hypothesis always assumes there is no correlation. It is important to note that $r$ is the PMCC from a sample of data – this is used as the test statistic – and $\rho$ (rho) is the PMCC for the population.

$H_0$ is always $H_0: \rho = 0$

$H_1$ can still be one-tailed or two-tailed:
$H_1: \rho < 0$ or $H_1: \rho > 0$ or $H_1: \rho \neq 0$

Test $r$ or use the table of critical values given on the formula sheet – these are all for one-tailed tests so halve the percentage for two-tailed tests. As usual, reject $H_0$ if $r$ lies in the critical region.

> **Example**
> Naomi thinks the number of ice creams sold on a beach are related to the temperature that day. She counts the number of ice creams sold on eight days and measures the temperature. She finds the PMCC is 0.73. Test, at the 10% significance level, to see whether the temperature and ice cream sales are correlated.

$H_0: \rho = 0$    $H_1: \rho \neq 0$ (two-tailed test)

Critical value from the table is 0.6215 for $n = 8$

| Product Moment Coefficient | | | | | |
|---|---|---|---|---|---|
| | Level | | | | Sample |
| 0.10 | 0.05 | 0.025 | 0.01 | 0.005 | Size, $n$ |
| 0.8000 | 0.9000 | 0.9500 | 0.9800 | 0.9900 | 4 |
| 0.6870 | 0.8054 | 0.8783 | 0.9343 | 0.9587 | 5 |
| 0.6084 | 0.7293 | 0.8114 | 0.8822 | 0.9172 | 6 |
| 0.5509 | 0.6694 | 0.7545 | 0.8329 | 0.8745 | 7 |
| 0.5067 | 0.6215 | 0.7067 | 0.7887 | 0.8343 | 8 |
| 0.4716 | 0.5822 | 0.6664 | 0.7498 | 0.7977 | 9 |
| 0.4428 | 0.5494 | 0.6319 | 0.7155 | 0.7646 | 10 |

Reject $H_0$ as 0.73 > 0.6215 and conclude there is evidence to suggest temperature and ice cream sales are correlated.

### Links to Other Concepts
- Hypothesis testing    ● Representing data
- Exponential equations
- Measures of location and spread (coding)

## QUICK TEST

1. The regression line of $y$ on $x$ is given as $y = 0.2 - 17x$. Find the regression line of $f$ on $g$ where $g = \frac{x}{4}$ and $y - 7 = f$.

2. The variables $x$ and $y$ are linked by the equation $y = ab^x$. Values of $\log y$ were plotted against $x$ and the regression line was $\log y = 0.35 - 0.12x$. Find $a$ and $b$.

3. Describe the correlation between $x$ and $y$ where:

   a)  $r = 0.867$          b)  $r = -0.482$

4. Calculate the PMCC for this data.

| $x$ | 20 | 30 | 42 | 48 | 52 | 63 |
|---|---|---|---|---|---|---|
| $y$ | 55 | 61 | 48 | 55 | 67 | 81 |

5. Use a 5% significance level to test whether $p$ and $q$ are negatively correlated:

| $p$ | 51 | 72 | 68 | 81 | 54 | 77 |
|---|---|---|---|---|---|---|
| $q$ | 58 | 44 | 50 | 38 | 60 | 46 |

6. Where $r = 0.5$, $n = 20$, test at the 1% significance level the hypotheses $H_0: \rho = 0$, $H_1: \rho > 0$.

## SUMMARY

- Rearrange equations and substitute to uncode a regression line.
- Take logs to uncode if variables are linked by exponential equations.
- PMCC is $r$ where $-1 \leq r \leq +1$:
  - $r = +1$, perfect positive correlation
  - $r = -1$, perfect negative correlation
  - $r = 0$, no correlation
- The PMCC is not affected by coding.
- $H_0$ is always $H_0: \rho = 0$
- Options for $H_1$ are $H_1: \rho < 0$ or $H_1: \rho > 0$ or $H_1: \rho \neq 0$.

## PRACTICE QUESTIONS

1. The growth of bacteria $y$ (in hundreds), each minute, $x$, can be modelled by the equation $y = ax^b$. A scatter diagram was plotted of $\log y$ against $\log x$ and the regression line was found to be $\log y = 0.78 + 2.62 \log x$.

   a)  Find $a$ and $b$, both to 3 significant figures. **[4 marks]**

   b)  Estimate how many bacteria there will be after 5 minutes. **[2 marks]**

2. Grapes are harvested from different vineyards and the tonnes per acre, $y$, are measured. The average number of hours per day of sunshine, $x$, for each vineyard is also recorded.

| Sunshine (hours) | 9 | 7 | 12 | 8 | 10 | 7.5 | 11.5 | 13 | 12.5 |
|---|---|---|---|---|---|---|---|---|---|
| Crop weight (tonnes) | 5 | 5.4 | 5.9 | 4.8 | 5.2 | 5.3 | 5.7 | 6 | 6.5 |

   a)  Explain what you understand by the PMCC. **[1 mark]**

   b)  Calculate the PMCC for this data. **[1 mark]**

   c)  Test at the 5% significance level whether there is a correlation between the weight of the grape crop and the number of hours of sunshine a vineyard had. State your hypotheses clearly. **[5 marks]**

# The Normal Distribution

The normal distribution is a **continuous** distribution (as opposed to the binomial, which is discrete). The normal distribution can take a full range of decimal values and models continuous data, such as heights.

## Properties of the Normal Distribution

The normal distribution is modelled by a bell-shaped curve and has these properties:

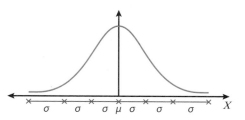

- It has two parameters, $\mu$ and $\sigma$ (the mean and the standard deviation).
- The distribution is symmetrical about the mean, $\mu$.
- The area under the curve is equal to 1.
- The mode, median and mean are all equal.
- About 68% of values lie within one standard deviation of the mean and 95% within two standard deviations of the mean.

A normal distribution is written as $X \sim N(\mu, \sigma^2)$, hence one with a mean of 60 and a standard deviation of 5 is written as $X \sim N(60, 5^2)$.

## Calculating Probabilities

Probabilities are calculated using A-level specific calculators. The cumulative probability function, CD, finds P(lower $< X <$ upper), or equally P(lower $\leq X \leq$ upper). Since the normal is a continuous distribution, the values for $\leq$ and $<$ are the same.

### Example

If $X \sim N(100, 25)$, input $\mu = 100$, $\sigma = 5$.

For P($X < 105$), input a very low value, e.g. −1000 for 'lower', as the distribution does go to −∞, and 105 as 'upper'. It is useful to visualise using a diagram.

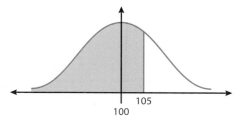

P($X < 105$) = 0.8413 (4 d.p.)

For P($X > 105$), input a very high value, e.g. 1000 for 'upper' and 105 as 'lower'.

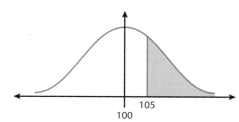

P($X > 105$) = 0.1587 (4 d.p.)

For P($90 < X < 105$), input 90 for 'lower' and 105 as 'upper'.

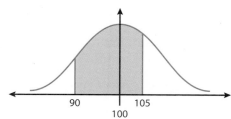

P($90 < X < 105$) = 0.8186 (4 d.p.)

The normal distribution models real-life scenarios.

## Example

The masses of lettuces sold at a market are normally distributed with mean 600 g and variance 400 g. If a lettuce is chosen at random, find the probability that its mass lies between 570 g and 610 g.

First write $X \sim N(600, 400)$.
Input into calculator: $\mu = 600$, $\sigma = 20$, lower 570, upper 610
Therefore $P(570 < X < 610) = 0.6247$ (4 d.p.)

## The Standard Normal Distribution

There are an infinite number of possibilities for the mean and standard deviation so, to standardise the distribution, a normal variable $Z$ is used. Any necessary probability for $X$ can be found by transforming $X$ into $Z$ using the formula $Z = \frac{X - \mu}{\sigma}$. The standard normal distribution is written as $Z \sim N(0, 1^2)$.

## The Inverse Normal Distribution

There are situations where probabilities are known but the $X$ value, the mean and/or the standard deviation are unknown.

If the $X$ value is unknown, it is found using the calculator's 'inverse normal' function. Input $\mu$ and $\sigma$ and the probability as the area (to the left on a diagram).

## Example

If $X \sim N(100, 25)$, find the value of $x$ if:

**a)** $P(X < x) = 0.8$

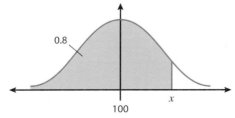

Input area = 0.8, $\mu = 100$, $\sigma = 5$

Calculator gives $x = 104.208$ (3 d.p.)

**b)** $P(X \geqslant x) = 0.43$

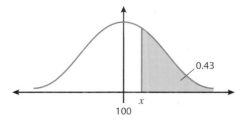

Input area = 0.57, $\mu = 100$, $\sigma = 5$

Calculator gives $x = 100.882$ (3 d.p.)

**c)** $P(x \leqslant X < 100) = 0.35$

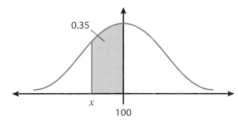

Input area = 0.15 (area to the left of $x$), $\mu = 100$, $\sigma = 5$

Calculator gives $x = 94.818$ (3 d.p.)

When $\mu$ or $\sigma$ are unknown, the probability is used to find the standard normal $Z$ value and then the formula $Z = \frac{X - \mu}{\sigma}$ is manipulated to find the unknown.

## Example

If $X \sim N(\mu, 30)$ and $P(X < 50) = 0.8624$, find the value of $\mu$.

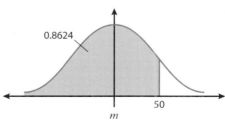

Use the calculator inverse function to find the $Z$ value with probability 0.8624.

Input area = 0.8624, $\mu = 0$, $\sigma = 1$, since $Z \sim N(0, 1^2)$

$$Z = 1.09 \text{ (2 d.p.)}$$

Now substitute into $Z = \frac{X - \mu}{\sigma}$, $1.09 = \frac{50 - \mu}{\sqrt{30}}$

Rearranging gives $\mu = 44.0$ (3 s.f.)

A similar method is used to find $\sigma$ if it is unknown. If both $\mu$ and $\sigma$ are unknown, they can be found by setting up and solving a pair of simultaneous equations. An example is shown below using the percentage points table.

## Percentage Points Table

The percentage points table is given for standard probability to $Z$ value conversions as an alternative to the calculator function. An extract from such a table is here:

| $p$ | $Z$ |
|--------|--------|
| 0.0500 | 1.6449 |
| 0.0250 | 1.9600 |
| 0.0100 | 2.3263 |

### Example

If $P(X < 45) = 0.05$ and $P(X > 80) = 0.01$, find the values of $\mu$ and $\sigma$.

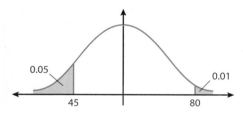

The $Z$ value for $p = 0.05$ is 1.6449, as seen in the table.

$x = 45$ is less than the mean, therefore this $Z$ value is negative. $Z = -1.6449$

The $Z$ value for $p = 0.01$ is 2.3263, as seen in the table. $x = 80$ is more than the mean, therefore this $Z$ value is positive. $Z = 2.3263$

Form two equations using $Z = \dfrac{X - \mu}{\sigma}$:

$$-1.6449 = \frac{45 - \mu}{\sigma} \qquad 2.3263 = \frac{80 - \mu}{\sigma}$$

$$-1.6449\sigma = 45 - \mu$$

$$2.3263\sigma = 80 - \mu$$

Solving simultaneously gives $\sigma = 8.81$, $\mu = 59.5$ (both to 3 s.f.)

**Links to Other Concepts**
- Simultaneous equations
- Measures of location and spread

## SUMMARY

- A normal distribution has two parameters, $\mu$ and $\sigma$. It is written as $X \sim N(\mu, \sigma^2)$.

- The distribution is symmetrical about the mean, $\mu$.

- The area under the curve is equal to 1.

- Mode = Median = Mean

- About 68% of values lie within one standard deviation of the mean.

  95% lie within two standard deviations of the mean.

  99.7% lie within three standard deviations of the mean.

- The standard normal distribution is written as $Z \sim N(0, 1^2)$.

- $Z = \dfrac{X - \mu}{\sigma}$ is used to transform $X$ values into $Z$ values.

1. If $X \sim N(500, 100)$, find:

   a) $P(X < 480)$

   b) $P(X \leqslant 480)$

   c) $P(X \geqslant 502)$

   d) $P(X < 508)$

2. If $X \sim N(60, 8^2)$, find:

   a) $P(58 < X < 64)$

   b) $P(55 < X \leqslant 57)$

3. If $X \sim N(625, 6^2)$, find $a$ such that:

   a) $P(X < a) = 0.75$

   b) $P(X \geqslant a) = 0.86$

   c) $P(625 < X < a) = 0.225$

4. Find the value of $\mu$ if $X \sim N(\mu, 100)$ and $P(X < 43) = 0.6241$

5. Find the value of $\sigma$ if $X \sim N(120, \sigma^2)$ and $P(X > 130) = 0.25$

6. Find the value of $\mu$ if $X \sim N(\mu, 12^2)$ and $P(X < 92) = 0.216$

7. Find the value of $\sigma$ if $X \sim N(20, \sigma^2)$ and $P(20 < X \leqslant 25) = 0.244$

8. If $P(X < 180) = 0.1$ and $P(X < 240) = 0.95$, find the values of $\mu$ and $\sigma$.

## PRACTICE QUESTIONS

1. The masses of apples collected from an orchard are normally distributed with a mean of 150 g and a standard deviation of 30 g.

   a) If an apple is chosen at random, find the probability that its mass lies between 130 g and 170 g. **[2 marks]**

   b) Find the mass exceeded by 6% of the apples. Give your answer to 3 significant figures. **[2 marks]**

   c) In one day 5000 apples are collected. Estimate how many will weigh more than 140 g. **[3 marks]**

2. Nails are produced with a mean length of 50 mm. Quality control checks show that 1.5% of nails are rejected because their length differs from the mean by at least 2 mm. Find the standard deviation of the length of the nails. **[5 marks]**

3. The lifetime of a particular brand of lightbulb is normally distributed with mean $\mu$ and standard deviation $\sigma$. 5% of the bulbs have a lifetime greater than 50 000 hours and 10% have a lifetime less than 25 000 hours. Find the mean and the standard deviation. **[5 marks]**

# The Normal Approximation and Hypothesis Testing

When $n$ is large, the binomial distribution can be reliably approximated by the normal distribution. This topic looks at the approximation of the binomial to the normal and at hypothesis testing with the normal distribution.

Read questions carefully and, if they have a fixed number of trials and a probability of success, use the binomial distribution (covered at AS-level). A mean and a variance (or standard deviation) is required for the normal distribution.

## Using the Normal Approximation to Approximate the Binomial Distribution

If the random variable $X$ is distributed binomially, i.e. $X \sim B(n, p)$, then it can be approximated by a normal distribution when $n$ is large and $p$ is approximately 0.5. It can be approximated by a normal distribution $Y$ where $Y \sim N(np, np(1-p))$.

An alternative way to check if the binomial can be approximated by the normal is to check that $np > 5$ and $n(1-p) > 5$.

## Continuity Correction

The binomial distribution is discrete and only takes integer values, whereas the normal distribution is continuous. This is accounted for by using a continuity correction. Think of the binomial probabilities as bars on a bar chart. Examples are:

| Binomial $X \sim B(n, p)$ | | Normal $Y \sim N(np, np(1-p))$ |
|---|---|---|
| $P(X < 4)$ | 3 4 5 | $P(Y < 3.5)$ |
| $P(X \leqslant 4)$ | 3 4 5 | $P(Y < 4.5)$ |
| $P(X > 4)$ | 3 4 5 | $P(Y > 4.5)$ |
| $P(X \geqslant 4)$ | 3 4 5 | $P(Y > 3.5)$ |

Algebraically this means:

| Binomial $X \sim B(n, p)$ | Normal $Y \sim N(np, np(1-p))$ |
|---|---|
| $P(X < x)$ | $P(Y < x - 0.5)$ |
| $P(X \leqslant x)$ | $P(Y < x + 0.5)$ |
| $P(X > x)$ | $P(Y > x + 0.5)$ |
| $P(X \geqslant x)$ | $P(Y > x - 0.5)$ |
| Also note: | |
| $P(X = x)$ | $P(x - 0.5 < Y < x + 0.5)$ |

## Example

If $X \sim B(150, 0.45)$, use an approximation to find $P(X < 70)$.

Find the mean: $\mu = np = 150 \times 0.45 = 67.5$

Find the standard deviation:

$\sigma = \sqrt{np(1-p)} = \sqrt{150 \times 0.45 \times 0.55}$

$= \sqrt{37.125} = 6.093$

Normal approximation: $Y \sim N(67.5, 37.125)$

Continuity correction: $P(X < 70) = P(Y < 69.5)$

Calculator gives: $= 0.6286$ (4 d.p.)

# Hypothesis Testing

The theory for hypothesis testing with the binomial distribution remains true. However, this time the parameter is $\mu$; the mean of a normal distribution is being tested.

## Don't forget

**Null hypothesis**: $H_0$, what we think is true, for normal written as $\mu = \cdots$

**Alternative hypothesis**: $H_1$, claims something other than the null hypothesis is true.

**One-tailed tests**: $H_1$, $\mu < \cdots$ or $\mu > \cdots$

**Two-tailed tests**: $H_1$, $\mu \neq \cdots$

**Significance level**: The probability of incorrectly rejecting $H_0$.

Remember that for a two-tailed test the significance level must be halved.

It is useful to look at critical regions diagrammatically. The significance level for these illustrations is 5%:

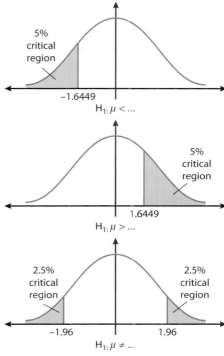

The sample mean $\bar{x}$ is used to find the test statistic for the normal distribution.

When $X \sim N(\mu, \sigma^2)$, then $\bar{X} \sim N\left(\mu, \dfrac{\sigma^2}{n}\right)$, where $n$ is the sample size.

The test statistic is $z = \dfrac{\bar{x} - \mu}{\dfrac{\sigma}{\sqrt{n}}}$

### Example

The times taken for swimmers at a local pool to swim a mile follow a normal distribution with mean 30 minutes and standard deviation 4. In the month before a charity swim, a group of swimmers increase their training sessions and in a random sample of 20 training sessions the mean had reduced to 29.1 minutes, while the variance didn't change. Test, at the 5% significance level, whether the mean time has reduced.

$H_0: \mu = 30$

$H_1: \mu < 30$ (one-tailed test)

$\bar{X} \sim N\left(30, \frac{4}{20}\right) \rightarrow \bar{X} \sim N(30, 0.2)$

There are two ways to look at this now:

### Method 1

$$z = \frac{29.1 - 30}{\sqrt{0.2}} = -2.012$$

From the percentage points table, 5% relates to 1.6449; as $H_1$ is <, the critical value is $-1.6449$.

Reject $H_0$ as $-2.012 < -1.6449$ and conclude there is evidence to suggest that the mean swimming time has reduced.

### Method 2

Find, using the normal distribution calculator function, $P(\bar{X} < 29.1) = 0.0221$ (4 d.p.)

Reject $H_0$ as $0.0221 < 0.05$ and conclude there is evidence to suggest that the mean swimming time has reduced.

## Links to Other Concepts
- The normal distribution
- Hypothesis testing
- The binomial distribution

## SUMMARY

- $X \sim B(n, p)$ can be approximated by $Y \sim N(np, np(1-p))$.

  - Used when $n$ is large and $p$ is approximately 0.5

  - Alternatively check that $np > 5$ and $n(1-p) > 5$

  - Use a continuity correction as going from discrete to continuous.

- When $X \sim N(\mu, \sigma^2)$, then $\bar{X} \sim N\left(\mu, \frac{\sigma^2}{n}\right)$, where $n$ is the sample size.

  - The test statistic is $z = \dfrac{\bar{x} - \mu}{\frac{\sigma}{\sqrt{n}}}$

## QUICK TEST

1. If $X \sim B(60, 0.48)$, use an approximation to find $P(X < 35)$.

2. If $X \sim B(80, 0.35)$, use an approximation to find $P(X \leqslant 37)$.

3. If $X \sim B(40, 0.43)$, use an approximation to find $P(X \geqslant 22)$.

4. If $X \sim B(95, 0.35)$, use an approximation to find $P(42 \leqslant X < 56)$.

5. $X \sim N(50, 10)$, $n = 5$. Test $\bar{x} = 52$ at the 10% significance level:

   $H_0: \mu = 50$

   $H_1: \mu > 50$

6. $X \sim N(85, 5^2)$, $n = 20$. Test $\bar{x} = 83$ at the 1% significance level:

   $H_0: \mu = 85$

   $H_1: \mu < 85$

7. $X \sim N(200, 16)$, $n = 20$. Test $\bar{x} = 202$ at the 5% significance level:

   $H_0: \mu = 200$

   $H_1: \mu \neq 200$

8. $X \sim N(60, 15)$, $n = 10$. Using the critical value, test $\bar{x} = 62$ at the 5% significance level:

   $H_0: \mu = 60$

   $H_1: \mu > 60$

## PRACTICE QUESTIONS

1. Plasterboard is cut into lengths by an industrial machine. The lengths of plasterboard are normally distributed with mean 240 cm and a standard deviation of 0.5 cm.

   a) Plasterboard lengths are classed as unsuitable if they are more than 1 cm shorter or longer than 240 cm. What is the probability that a length of plasterboard is unsuitable? **[2 marks]**

   Ten lengths of plasterboard are bundled together to be sold as a pack.

   b) What is the probability that none of the lengths in a pack are unsuitable? **[2 marks]**

   100 packs are loaded onto a lorry for delivery to a DIY store.

   c) Use a suitable approximation to find the probability that at least 70 of the packs contain no unsuitable lengths. **[4 marks]**

2. Batteries are found to have a mean lifetime of 45 hours and a standard deviation of 4.4 hours. A factory foreman tested 30 batteries one week and found the mean lifetime to be 44 hours. Assume the standard deviation is unchanged. Test, at the 5% significance level, whether this shows a change in the mean lifetime of the batteries. **[5 marks]**

# Probability

Probability is extended at A-level to include conditional probability. The use of statistical notation is expected in answering questions. Knowledge of Venn diagrams and tree diagrams is assumed from AS-level.

## Two-way Tables

Information may be presented in a two-way table and probability questions asked. Totals may be given but usually they will need to be calculated.

### Example

Information about children in a Year 1 class is collected and shown in this table:

|  | Glasses (G) | Not glasses (G′) |
|---|---|---|
| Blue eyes (Bl) | 3 | 9 |
| Brown eyes (Br) | 2 | 4 |
| Other (Ot) | 5 | 7 |

If one child is chosen at random, find:

**a)** $P(Ot)$

Completing the totals on the two-way table:

|  | Glasses (G) | Not glasses (G′) | Total |
|---|---|---|---|
| Blue eyes (Bl) | 3 | 9 | 12 |
| Brown eyes (Br) | 2 | 4 | 6 |
| Other (Ot) | 5 | 7 | 12 |
| Total | 10 | 20 | 30 |

$$P(Ot) = \frac{12}{30} = \frac{2}{5}$$

**b)** $P(G \cap Bl)$

$$P(G \cap Bl) = \frac{3}{30} = \frac{1}{10}$$

**c)** $P(Br \cup G')$

$$P(Br \cup G') = \frac{22}{30} = \frac{11}{15}$$

## Conditional Probability

For two events, A and B, the probability of A happening, given that B has already happened, is written as $P(A \mid B)$. This is calculated using a formula that must be known:

$$P(A \mid B) = \frac{P(A \cap B)}{P(B)}$$

### Example

If $P(A) = \frac{1}{5}$, $P(B) = \frac{1}{3}$ and $P(A \mid B) = \frac{3}{7}$, find $P(A \cap B)$.

Rearrange $P(A \mid B) = \frac{P(A \cap B)}{P(B)}$ to give:

$$P(A \cap B) = P(A \mid B) \times P(B)$$

$$P(A \cap B) = \frac{3}{7} \times \frac{1}{3} = \frac{1}{7}$$

Remember there are other formulae that may help:

$$P(A') = 1 - P(A)$$

$$P(A \cup B) = P(A) + P(B) - P(A \cap B)$$

## Tree Diagrams (without replacement)

Tree diagrams can be used to show probabilities for events without replacement.

### Example

There are four red sweets and six yellow sweets in a bag. Two sweets are chosen at random. The first sweet is eaten, then the second sweet is taken.

**a)** Find the probability that both sweets were red.

Notice on the tree diagram that on the second set of branches there is one fewer sweet in total as the first sweet wasn't replaced.

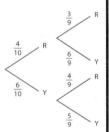

$$P(R \cap R) = \frac{4}{10} \times \frac{3}{9} = \frac{2}{15}$$

**b)** Find the probability that the second sweet was red given that the first sweet was yellow.

On the diagram, the second sweet (red) is denoted $R_2$ and the first sweet (yellow) is denoted $Y_1$.

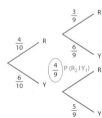

Using the formula:

$$P(R_2 \mid Y_1) = \frac{P(R_2 \cap Y_1)}{P(Y_1)} = \frac{\frac{6}{10} \times \frac{4}{9}}{\frac{6}{10}} = \frac{4}{9}$$

## Assumptions

Many assumptions are made when dealing with probabilities. Events are described as being **random** but, realistically, are they random? Is the experiment **unbiased**? Are the probabilities really **equally likely**?

It is an important skill to be able to critique assumptions made in questions and to be able to suggest assumptions that may be more realistic. One of the most common incorrect assumptions is to assume that two events are independent. Remember that for two events to be independent then
**P(A ∩ B) = P(A) × P(B)**.

### Example

Abdi wanted to find the probability that the college gym would be full at lunchtime on Tuesday. He counted the number of students using the gym at lunchtime during a particular week and it was full two days out of five. Abdi says that the probability it will be full on Tuesday is $\frac{2}{5}$. Comment on Abdi's assumption.

$\frac{2}{5}$ is the relative frequency over five days. Five is not a large enough sample of days. Additionally, the use of the gym on each day may not be independent.

### Links to Other Concepts
● Venn diagrams ● Tree diagrams

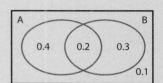
## PRACTICE QUESTIONS

1. Two events, R and Q, are such that $P(R) = \frac{8}{15}$, $P(Q) = q$, $P(R \cap Q) = \frac{1}{3}$ and $P(R \mid Q) = \frac{4}{7}$.

   **a)** Find the value of $q$. [2 marks]

   **b)** Find P(Q | R′). [3 marks]

   **c)** Determine whether R and Q are independent. [3 marks]

2. Cards labelled A and B and C are in a closed box. There are 12 cards labelled A, 7 cards labelled B, and 1 card labelled C. Two cards are chosen at random. The first card is not replaced before the second one is chosen.

   **a)** Draw a tree diagram to represent the possibilities for the cards chosen. [3 marks]

   **b)** Find the probability that the cards chosen are the same. [2 marks]

   **c)** Given the first card chosen was labelled B, find the probability that the second card is labelled C. [2 marks]

# Answers

## Indices and Surds

### QUICK TEST (page 7)

1. a) $q^{\frac{3}{4}}$  b) $q^{-2}$  c) $q^{\left(\frac{3}{4}+-2\right)} = q^{-\frac{5}{4}}$  d) $q^{-12}$
2. C
3. $\frac{1}{x}$
4. a) $\frac{2\sqrt{3}}{3}$  b) $\frac{\sqrt{15}+6}{-7} = \frac{-\sqrt{15}-6}{7}$
5. a) $2^{\frac{3}{2}}$  b) $3a^{-1} \times 2^{-\frac{1}{3}}$
6. Scale factor $= \frac{1}{\sqrt{7}} = \frac{\sqrt{7}}{7}$

### PRACTICE QUESTIONS (page 7)

1. a) $\frac{3\sqrt{a}}{3\sqrt{b}-\sqrt{a}} \times \frac{3\sqrt{b}+\sqrt{a}}{3\sqrt{b}+\sqrt{a}}$ **[1]**

   $\frac{9\sqrt{ab}+3a}{9b-a}$ **[1]**

   b) $\frac{9\sqrt{ab}+3a}{9b-a} = \frac{18\sqrt{5}+15}{31}$

   For setting up at least a pair of simultaneous equations by comparing coefficients from rationalised expression in part **a)** **[1]**

   $3a = 15 \rightarrow a = 5$

   $9\sqrt{ab} = 18\sqrt{5}$

   $81ab = 1620$

   $ab = 20 \rightarrow b = 4$

   $9b - a = 31$   checking   $9 \times 4 - 5 = 31$ **[2]**

2. $(2^a \cos(30))^{-\frac{4}{a}} = (2^a)^{-\frac{4}{a}} \times \left(\frac{\sqrt{3}}{2}\right)^{-\frac{4}{a}}$ **[1]**

   $= (2)^{a \times -\frac{4}{a}} \times \left(\frac{2}{\sqrt{3}}\right)^{\frac{4}{a}}$ **[1]**

   $= 2^{-4} \times \left(\frac{16}{9}\right)^{\frac{1}{a}}$ **[1]**

   $= \frac{1}{16} \times \sqrt[a]{\frac{16}{9}} = \sqrt[a]{\left(\frac{1}{16}\right)^a \times \frac{16}{9}}$ **[1]**

   $= \sqrt[a]{\frac{1}{9 \times 16^{a-1}}}$ **[1]**

3. a) $\frac{-3p \pm \sqrt{9p^2 - 4p^2}}{2p} = \frac{-3p \pm p\sqrt{5}}{2p}$ **[1]**

   $x = \frac{-3+\sqrt{5}}{2}$ and $x = \frac{-3-\sqrt{5}}{2}$ **[1]**

   b) $\frac{1}{x} = \frac{2}{-3+\sqrt{5}}$

   $= \frac{2}{-3+\sqrt{5}} \times \frac{-3-\sqrt{5}}{-3-\sqrt{5}}$ **[1]**

   $= \frac{(-6-2\sqrt{5})}{9-5} = \frac{-3-\sqrt{5}}{2}$ **[1]**

   $\frac{1}{x} = \frac{2}{-3-\sqrt{5}}$

   $= \frac{2}{-3-\sqrt{5}} \times \frac{-3+\sqrt{5}}{-3+\sqrt{5}}$ **[1]**

   $= \frac{-6+2\sqrt{5}}{9-5} = \frac{-3+\sqrt{5}}{2}$ **[1]**

## Polynomials, Quadratics and Simultaneous Equations

### QUICK TEST (page 11)

1. $(-1, -5)$ and $(6, 2)$
2. $f(x) = (2x + 1)(x - 3)(x + 2)$

1. a)

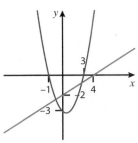

   For straight line with positive gradient passing through $y$-axis below the origin **[1]**

   For curve in correct orientation with two distinct $x$-intercepts **[1]**

   For attempt to find $x$-intercepts for quadratic either by factorisation or other quadratic method $(x + 1)(x - 3)$ **[1]**

   For all axis-intercepts labelled correctly **[1]**

   For two clear intercepts of the two lines identified (both intercepts are below the $x$-axis so have a negative $y$-coordinate) **[1]**

   b) For method to solve as simultaneous equations (substitution or elimination both valid) **[1]**

   $x^2 - 2x - 3 = \frac{1}{2}x - 2 \rightarrow 2x^2 - 4x - 6 = x - 4$

   $\rightarrow 2x^2 - 5x - 2 = 0$ **[1]**

2. a) By factor theorem $6(-2)^3 + 20(-2)^2 + 17(-2) + 2 = 0$, therefore $(x + 2)$ is a factor and $x = -2$ a possible solution. **[1]**

   For a method of finding the factors, e.g. comparing coefficients: $px^3 + (2p + q)x^2 + (2q + r)x + 2r = 6x^3 + 20x^2 + 17x + 2$ **[1]**

   $p = 6, q = 8, r = 1$ **[2]**

   b) $x^2 + \frac{4}{3}x + \frac{1}{6} = 0$ **[1]**

   $\left(x + \frac{2}{3}\right)^2 - \left(\frac{2}{3}\right)^2 + \frac{1}{6} = 0$ **[1]**

   $\left(x + \frac{2}{3}\right)^2 = \frac{5}{18}$

   $x + \frac{2}{3} = \pm\sqrt{\frac{5}{18}}$ **[1]**

   $x = -\frac{2}{3} \pm \sqrt{\frac{5}{18}} = -\frac{2}{3} \pm \frac{\sqrt{5}}{3\sqrt{2}} = -\frac{4}{6} \pm \frac{\sqrt{5 \times 2}}{6}$ **[1]**

   $= \frac{-4 \pm \sqrt{10}}{6}$ **[1]**

   c) Rearrange to form $6a^{1.2} + 8a^{0.6} + 1 = 0$ **[1]**

   From part **a)**, $a^{0.6} = \frac{-4+\sqrt{10}}{6}$ or $a^{0.6} = \frac{-4-\sqrt{10}}{6}$ **[1]**

   For a method of finding $a$. (Alternate method to taking the 0.6-root is to raise both sides to power $\left(\frac{5}{3}\right)$) **[1]**

   $a^{0.6} = \frac{-4+\sqrt{10}}{6} \rightarrow a = \sqrt[0.6]{\frac{-4+\sqrt{10}}{6}} \rightarrow a = -0.0376$ **[1]**

   $a^{0.6} = \frac{-4-\sqrt{10}}{6} \rightarrow a = \sqrt[0.6]{\frac{-4-\sqrt{10}}{6}} \rightarrow a = -1.34$ **[1]**

## Inequalities and Partial Fractions

### QUICK TEST (page 15)

1.

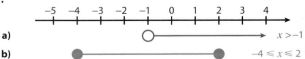

   a) $x > -1$

   b) $-4 \leqslant x \leqslant 2$

   c) $-1 < x \leqslant 2$

2. $\frac{A}{5x+1} + \frac{B}{x-3} + \frac{C}{(x-3)^2}$

## PRACTICE QUESTIONS (page 15)

**1.** An attempt at any valid method **[1]**

When $x = 1$: $\qquad 10 = 24C$

$\qquad\qquad C = \frac{5}{12}$ **[1]**

When $x = -\frac{1}{2}$: $\qquad -8 = -\frac{21}{4}B$

$\qquad\qquad B = \frac{32}{21}$ **[1]**

When $x = -\frac{5}{3}$: $\qquad -22 = \frac{56}{9}A$

$\qquad\qquad A = -\frac{99}{28}$ **[1]**

**2.**

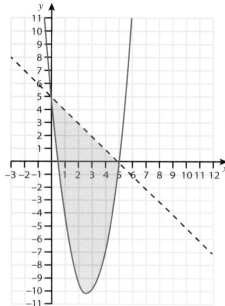

For correct line, dotted with shading only below **[1]**
For quadratic curve with correct $x$-intercepts and $y$-intercept **[1]**
For minimum point on graph **[1]**
For correct region shaded on the graph **[1]**

**3. a)** $f\left(-\frac{1}{3}\right) = 6\left(-\frac{1}{3}\right)^2 - 13\left(-\frac{1}{3}\right) - 5$ **[1]**

$\qquad = \frac{2}{3} + \frac{13}{3} - 5 = 0$

By factor theorem $(3x + 1)$ is a factor of $6x^2 - 13x - 5$ **[1]**

**b)** $\dfrac{34x^2 - 68x}{12x^3 - 26x^2 - 10x} = \dfrac{2x(17x - 34)}{2x(6x^2 - 13x - 5)}$ **[2]**

$\qquad = \dfrac{17x - 34}{6x^2 - 13x - 5}$ **[1]**

**c)** $\dfrac{34x^2 - 68x}{12x^3 - 26x^2 - 10x} = \dfrac{17x - 34}{6x^2 - 13x - 5} = \dfrac{17x - 34}{(3x + 1)(2x - 5)}$ **[1]**

For identifying form as $\dfrac{A}{3x+1} + \dfrac{B}{2x-5}$ **[1]**
For any valid method used to find $A$ and $B$, e.g.
$\dfrac{17x - 34}{(3x + 1)(2x - 5)} = \dfrac{A}{3x+1} + \dfrac{B}{2x-5}$
$17x - 34 = A(2x - 5) + B(3x + 1)$

$17 = 2A + 3B$
$-34 = -5A + B \qquad\qquad B = 5A - 34$ **[1]**
$17 = 2A + 3(5A - 34)$

$17A = 17 + 102 = 119$ **[1]**
$A = 7$ **[1]**
$B = 1$ **[1]**

**4. Possible algebraic solution:**

$ax - b < 0$ gives the result $x < 7$, only inequality in given solution that is a $<$ (rather than a $\leqslant$). **[1]**

$ax < b \qquad \rightarrow \qquad x < \frac{b}{a} \qquad \rightarrow \qquad 7 = \frac{b}{a}$ **[1]**

Positive bound from quadratic is from $\left(x - \frac{b}{a+5}\right) = 0$, as $a$ and $b$ are both positive.

$x = \frac{b}{a+5} \qquad \rightarrow \qquad \frac{b}{a+5} = 2$ **[2]**

Rearrange one equation to isolate unknown $b = 7a$
Substitute into second equation $\frac{7a}{a+5} = 2$ **[1]**
Rearrange and solve for one unknown:

$7a = 2a + 10 \qquad \rightarrow \qquad 5a = 10 \qquad \rightarrow \qquad a = 2$ **[1]**
Find second unknown: $b = 7 \times 2 = 14$ **[1]**

**Possible diagrammatic solution:**

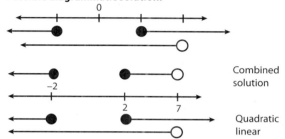

Combined solution

Quadratic linear

**[2]**

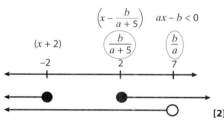

**[2]**

$\frac{b}{a+5} = 2$ and $\frac{b}{a} = 7$

Rearrange one equation to isolate unknown: $b = 7a$ **[1]**
Substitute into second equation: $\frac{7a}{a+5} = 2$ **[1]**
Rearrange and solve for one unknown:
$7a = 2a + 10 \qquad \rightarrow \qquad 5a = 10 \qquad \rightarrow \qquad a = 2, b = 14$ **[1]**

## Functions, Graphs and Transformations

### QUICK TEST (page 21)

**1. a)** $y = f(-2x)$ $\qquad\qquad$ **b)** $y = f(x + 5)$

**2.** $-5$ and $3$

**3.**

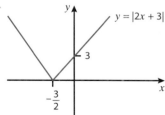

**4.**

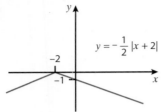

$$y = -\frac{1}{2}|x + 2|$$

**5.** Range $-3 \leqslant y < 45$

**6. a)** $ff(x) = 9x + 4$      **b)** $gf(x) = 9x^2 + 6x - 1$

**c)**

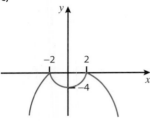

**d)** $f^{-1}(x) = \frac{x-1}{3}$

## PRACTICE QUESTIONS (page 21)

**1. a)**

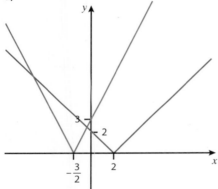

$-x + 2 = 2x + 3$ **[1]**

$3x = -1$ **[1]**

$x = -\frac{1}{3}$ **[1]**

$y = \frac{7}{3}$ **[1]**

$-x + 2 = -2x - 3$ **[1]**

$x = -5, \, y = 7$ **[1]**

**b)** $-5 < x < -\frac{1}{3}$ **[2]**

**2. a)** $f(x)$ has the effect of a translation of $\begin{pmatrix} 0 \\ -2 \end{pmatrix}$ so $g(x)$ must create the

stretch and a translation of $\begin{pmatrix} 0 \\ -1 \end{pmatrix}$ **[1]**

(as the operations are in perpendicular directions they are
independent in how they are applied)

$g(x^2) = \left(\frac{1}{2}x\right)^2 - 1$ **[1]**

$= \frac{1}{4}x^2 - 1$ **[1]**

So $g(x) = \frac{1}{4}x - 1$ **[1]**

**b) Either:**

At $x = 2$, $y = x^2 = 4$

Horizontal stretch means for $y = fg(x^2)$ at $x = 4$, $y = 4$ **[1]**

Vertical translation $-3$ **[1]**

So $y$-coordinate is 1 **[1]**

**Or:**

$y = fg(x^2) = \frac{1}{4}x^2 - 3$ **[2]**

At $x = 4$, $y = \frac{1}{4}4^2 - 3 = \frac{16}{4} - 3 = 1$ **[1]**

# Day 2
## Coordinate Geometry
### QUICK TEST (page 25)

**1.** A–G; B–E; C–F; D–H

**2.** Midpoint $\left(-\frac{1}{2}, 1\right)$; Length 13

**3.** Gradient of normal $= \frac{(12 + 2\sqrt{10}) - 12}{-1 - -10} = \frac{2\sqrt{10}}{9}$

Gradient of tangent $= \frac{-9}{2\sqrt{10}} = \frac{-9\sqrt{10}}{20}$

### PRACTICE QUESTIONS (page 25)

**1. a)** $x = \tan\theta - 3$     $\rightarrow$     $\tan\theta = x + 3$ **[1]**

$y = \tan 2\theta = \frac{2\tan\theta}{1 - \tan^2\theta}$ **[1]**

$y = \frac{2(x + 3)}{1 - (x + 3)^2}$ **[1]**

$y = \frac{2x + 6}{1 - (x^2 + 6x + 9)}$

$y = -\frac{2x + 6}{x^2 + 6x + 8}$ **[1]**

$y = -\frac{2x + 6}{(x + 2)(x + 4)}$ **[1]**

**b)** Asymptotes are at $x = -2$ and $x = -4$ **[1]**

$y$-intercept when $x = 0$

$y = -\frac{6}{8}$ so coordinate $\left(0, -\frac{3}{4}\right)$ **[1]**

$x$-intercept when $y = 0$

$2x + 6 = 0$     $\rightarrow$     $x = -3$ **[1]**

**2.** Gradient of tangent $= -\frac{4}{3}$ **[1]**

The radius is a section of the line with equation $y - 5 = \frac{3}{4}(x - 1)$ **[1]**

Let $(p, q)$ be a point on the line $y - 5 = \frac{3}{4}(x - 1)$ which is 5 units
from $(1, 5)$

Gives equations:   $5 = \sqrt{(p - 1)^2 + (q - 5)^2}$ and $q - 5 = \frac{3}{4}(p - 1)$ **[1]**

Method of solving by simultaneous equations:

$25 = (p - 1)^2 + \left(\frac{3}{4}(p - 1)\right)^2$

$25 = \frac{25}{16}(p - 1)^2$

$16 = (p - 1)^2$

$p - 1 = \pm 4$

$p = -3$ or $p = 5$ **[2]**

$q = 2$ or $q = 8$ **[1]**

Equation of circle is either:

$(y - 2)^2 + (x + 3)^2 = 25$ or $(y - 8)^2 + (x - 5)^2 = 25$ **[2]**

## Trigonometry 1
### QUICK TEST (page 31)

**1. a)** $\frac{2\pi}{15}$ radians    **b)** $144°$    **c)** $7\pi$ radians    **d)** $60.7°$ (3 s.f.)

**2.** 15.2 (3 s.f.)

**3.** 29.9° (3 s.f.) or $\frac{\pi}{6}$ radians

**4. a)** Area $= \frac{26\pi}{7}$ cm²    **b)** Perimeter $= (2 \times 2) + \left(2 \times \frac{13\pi}{7}\right) = 4 + \frac{26\pi}{7}$ cm

**5.** $\frac{m\sqrt{3}-2}{4}$

**6. a)** $\sin 7.5° = \sin\frac{\pi}{24} \approx \frac{\pi}{24}$

**b)** $\cos 0.02 - \tan 0.01 \approx 1 - \frac{1}{2} \times 0.02^2 - 0.01 \approx 0.9898$

**c)** 0.286% (3 s.f.)

**7. a)** $\frac{\sqrt{2}}{2}$    **b)** $\frac{1}{2}$    **c)** 1    **d)** $\sqrt{3}$    **e)** $\frac{2}{3}$

## PRACTICE QUESTIONS (page 31)

**1.** $\tan x = 0$ or $3\cos x - 1 = 0$ or $5\sin x - 2 = 0$ **[1]**

$\tan x = 0$       $x = -\pi, 0, \pi, 2\pi$ **[1]**

$3\cos x - 1 = 0$    $\cos x = \frac{1}{3}$    $x = -1.23, 5.05$ **[1]**

$5\sin x - 2 = 0$    $\sin x = \frac{2}{5}$    $x = 0.41, 2.73$ **[1]**

(all answers to 2 d.p.)

**2.** Triangle ABE: apply area rule $\left(A = \frac{1}{2}ab\sin C\right)$ to find AE

$AE = \frac{2 \times 45}{9\sin 45} = 10\sqrt{2}$ **[1]**

Use cosine rule to find BE:

$BE = \sqrt{(10\sqrt{2})^2 + 9^2 - 2 \times 10\sqrt{2} \times 9 \times \cos 45} = \sqrt{101}$ **[1]**

Sine rule to find BD:

$BD = \frac{\sqrt{101} \times \sin 30}{\sin 45} = \frac{\sqrt{202}}{2}$ **[1]**

Angle $EBD = 180 - 45 - 30 = 105°$ **[1]**

Area of $BDE = \frac{1}{2} \times BE \times BD \times \sin 105 = 34.4921314...$ **[1]**

Area of $BCD = \frac{1}{2} \times 6 \times BD = \frac{3\sqrt{202}}{2}$ **[1]**

Total area $= 45 + 34.492131... + \frac{3\sqrt{202}}{2} = 100.8111...$

$= 101\,cm^2$ (3 s.f.) **[1]**

## Trigonometry 2

### QUICK TEST (page 35)

**1.** $\tan(A + B) = \frac{\tan A + \tan B}{1 - \tan A \tan B}$

$\tan(2p) = \tan(p + p) = \frac{\tan p + \tan p}{1 - \tan p \tan p} = \frac{2\tan p}{1 - \tan^2 p}$

**2.** $\frac{\pi}{2}, \frac{7\pi}{6}, \frac{3\pi}{2}, \frac{11\pi}{6}$

**3.** A–J, B–L, C–I, D–G, E–H, F–K

### PRACTICE QUESTIONS (page 35)

**1. a)** For converting $\tan^2 x$ into $\sec^2 x$ using identity

$1 + \tan^2 x = \sec^2 x$: **[2]**

$\tan^2 x + \frac{3}{1 - 2\sin^2\frac{x}{2}} - 9 \equiv 1 + \tan^2 x + \frac{3}{1 - 2\sin^2\frac{x}{2}} - 10$

$\equiv \sec^2 x + \frac{3}{1 - 2\sin^2\frac{x}{2}} - 10$

Use double angle formulae (in sine, or in cosine if denominator converted into cosine):

$\equiv \sec^2 x + \frac{3}{\cos x} - 10$ **[2]**

$\equiv \sec^2 x + 3\sec x - 10$ **[1]**

**b)** For method of solving quadratic: **[1]**

$(\sec x - 2)(\sec x + 5) = 0$

$\sec x = 2$       $\cos x = \frac{1}{2}$

$x = \frac{\pi}{3}, \frac{5\pi}{3}$ **[1]**

$\sec x = -5$       $\cos x = -\frac{1}{5}$

$x = 1.77, 4.51$ (3 s.f.) **[1]**

For full set of correct answers to suitable degree of accuracy. **[1]**

**2. a)** $10\cos\alpha = 6$ **[1]**

$\cos\alpha = 0.6$ **[1]**

$\alpha = 53.130...° = 53.1°$ (3 s.f.) **[1]**

$p = 10\sin\alpha = 10\sin 53.130... = 8$ **[1]**

**b)** Asymptotes are present when the denominator $= 0$ **[1]**

Asymptotes present when $6\sin x - 8\cos x = -q$

$-q = 10\sin(\theta - \alpha)$ **[1]**

If $\sin(\theta - \alpha) = -\frac{q}{10}$ then the denominator is 0 and an asymptote is formed.

The horizontal transformation, created by the $-\alpha$, makes no difference vertically. So $-1 \leqslant -\frac{q}{10} \leqslant 1$. **[1]**

For no asymptotes, $q$ must satisfy the opposite conditions:

$q < -10$ or $q > 10$ **[2]**

**c)** Consider the curve

$y = 6\sin x - 8\cos x + 12 = 10\sin(\theta - 53.130...) + 12$ **[1]**

Maximum 22 and minimum 2 **[1]**

So $y = \frac{1}{6\sin x - p\cos x + 12}$ maximum at $\frac{1}{2}$ **[1]**

And minimum is $\frac{1}{22}$ **[1]**

## Trigonometry 3

### QUICK TEST (page 37)

**1.** 8

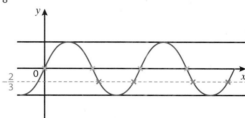

**2. a)** $x = 75$ ✓, 110 ✗, 160 ✗, 195 ✓, 255 ✓

15, 105, 165, 285, 345

**b)** $x = 15$ ✗, 95 ✗, 185 ✗, 215 ✓

35

**c)** $x = 30$ ✓, 60 ✓, 172.5 ✗, 210 ✓, 240 ✓

120, 150, 300, 330

### PRACTICE QUESTIONS (page 37)

**1.** Either $\tan^2\left(\frac{\theta}{3}\right) = 25$   or   $5\sin\frac{\theta}{2} = 0$ **[1]**

$\tan\left(\frac{\theta}{3}\right) = \pm 5$

$\left(\frac{\theta}{3}\right) = \tan^{-1}(5) = 78.6900...$     $\rightarrow$     $\theta = 236.07020...$ **[1]**

$\left(\frac{\theta}{3}\right) = \tan^{-1}(-5) = -78.6900...$ (can be found by rotational symmetry rather than arctan)     $\rightarrow$     $\theta = -236.07020...$ **[1]**

All results in the range gained from $y = \tan\left(\frac{\theta}{3}\right)$ having a period of 540

−843.9, −303.9, **236.1**, 776.1

−776.1, **−236.1**, 303.9, 843.9 **[2]**

$\frac{\theta}{2} = \sin^{-1}0 = 0$     $\rightarrow \theta = 0$ **[1]**

Graph of $y = \sin x$ crosses the $x$-axis every 180°, so $y = \sin\frac{x}{2}$ will cross every 360°

−720, −360, 0, 360, 720 **[1]**

**2. a)** $3(1 - \cos^2\alpha) = 2 - 2\cos\alpha$ **[1]**

$3\cos^2\alpha - 2\cos\alpha - 1 = 0$

$(3\cos\alpha + 1)(\cos\alpha - 1) = 0$ **[1]**

$\cos\alpha = -\frac{1}{3}$   and   $\cos\alpha = 1$ **[1]**

$\alpha = \arccos-\frac{1}{3} = 109.47\ldots \rightarrow -109.47\ldots, 109.47\ldots, 250.52\ldots$ **[2]**

$\alpha = \arccos 1 = 0 \rightarrow 0, 360$ **[2]**

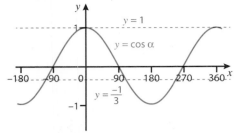

**b)** $3\sin^2(2x) = 2(1 - \cos 2x)$

$2x = \arccos\left(-\frac{1}{3}\right)$ and $2x = \arccos 1$ **[1]**

$x = \frac{1}{2}\arccos-\frac{1}{3} = 54.7356\ldots$

$\rightarrow -125.26\ldots, -54.73\ldots, 54.73\ldots, 125.26\ldots, 305.26\ldots, 234.73\ldots$ **[2]**

$x = \frac{1}{2}\arccos 1 = 0$

$\rightarrow -180, 0, 180, 360$ **[1]**

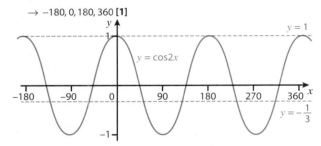

# Day 3
## Binomial Expansion
### QUICK TEST (page 41)

**1.** For the expansion to be valid, $-4 < x < 4$. For $4 + x = 2.3$, $x = -1.7$, which is within the range, so yes.

### PRACTICE QUESTIONS (page 41)

**1. a)** $2^p = \sqrt[3]{128} = \sqrt[3]{2^7}$ **[1]**

$p = \frac{7}{3}$ **[1]**

**b)** $14\sqrt[3]{2} = \sqrt[3]{128} \times \frac{7}{3} \times \frac{b}{2}$ **[1]**

$b = 3$ **[1]**

**c)** $-\frac{2}{3} < x < \frac{2}{3}$ **[1]**

**d)** $(2 + 3x)^{\frac{7}{3}} = (2.06)^{\frac{7}{3}}$ **[1]**

Use $x = 0.02$ **[1]**

$(2.06)^{\frac{7}{3}} \approx \sqrt[3]{128} + 14\sqrt[3]{2} \times 0.02 + \frac{7\sqrt[3]{2}}{2} \times 0.02^2$ **[1]**

$(2.06)^{\frac{7}{3}} \approx 5.0396841995\ldots + 0.35277789397\ldots + 0.0017638894698\ldots$ **[1]**

$\approx 5.394$ (4 s.f.) **[1]**

## Sequences and Series
### QUICK TEST (page 47)

**1. a)** Geometric, increasing, 48.6
**b)** Neither, periodic, $\frac{7}{4}$
**c)** Geometric, decreasing, 2048, 2
**d)** Neither, decreasing to a limit, $u_1 = 19, u_8 = 1.2502272, L = 1.25$

**2.** $1, 1, 2, 3, 5, 8, 13, \ldots$

**3.** Note: without a term number being defined, there are infinite sequences the term-to-term rule could represent.

**a)** $u_{n+1} = u_n - 3, u_1 = 15$    $u_n = 18 - 3n$    $u_{100} = -282$
**b)** $u_{n+1} = 4u_n, u_1 = 0.2$    $u_n = 0.2 \times 4^{n-1}$    $u_8 = 3276.8$
**4. a)** $r = 0.2$    **b)** $a = 10000$    **c)** $3.99872$
**5.** 66
**6. a)** $L = \frac{10a}{7}$    **b)** $a = 1.4 = \frac{7}{5}$

### PRACTICE QUESTIONS (page 47)

**1. a)** $u_5 = a + 4d = q^2 - 2$ **[1]**

$u_3 = a + 2d = 3q$ **[1]**

$u_5 - u_3 = 2d = q^2 - 3q - 2$ **[1]**

**b)** $a + 2d = a + (q^2 - 3q - 2) = 3q$ **[1]**

$a = -q^2 + 6q + 2$ **[1]**

**c)** $S_5 = 75 = (a) + (a + d) + (a + 2d) + (a + 3d) + (a + 4d)$

$= 5a + 10d = 5(a + 2d)$ **[1]**

$75 = 5(a + 2d)$ and $a + 2d = 3q$ **[1]**

$75 = 5(3q)$

$3q = 15$ **[1]**

$q = 5$ **[1]**

**2. a)** $u_1 - v_1 = 3193.6$     For statement or implied use. **[1]**

Substitute in $u_1 = 500v_1$

$499v_1 = 3193.6$ **[1]**

$v_1 = 6.4$ **[1]**

$u_1 = 3193.6 + v_1 = 3200$ **[1]**

**b)** For geometric series $S_\infty = \frac{a}{1-r}, |r| < 1$

$a = 3200$ **[1]**

$3200 + u_1 = S_\infty$ **[1]**

$\frac{3200}{1-r} = 6400$ **[1]**

$6400r = 6400 - 3200$

$r = \frac{1}{2} = 0.5$ **[1]**

**c)** Start by considering the arithmetic sequence to find the common difference, $d$:

$v_1 = a = 6.4$

$v_2 = u_2 - w_2$ **[1]**

$= (3200 \times 0.5) - 1593.5 = 6.5$ **[1]**

$d = v_2 - v_1$ **[1]**

$= 0.1$ **[1]**

$\sum_{n=1}^{10}(w_n) = \sum_{n=1}^{10}(u_n) - \sum_{n=1}^{10}(v_n)$ **[1]**

$= \frac{3200(1 - 0.5^{10})}{1 - 0.5} - \frac{10}{2}(2 \times 6.4 + (10 - 1) \times 0.1)$ **[1]**

$= 6393.75 - 68.5 = 6325.25$ **[1]**

## Exponentials
### QUICK TEST (page 49)

**1. a)** $y = \frac{3}{e}$

**b)** $x = 1$

**2. a)** $53000$

**b)** $P = \frac{53000}{e^{10}} = 2$ (nearest whole number)

### PRACTICE QUESTIONS (page 49)

**1. a)**

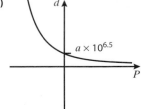

For shape of curve [1]
For $y$-axis intercept [1]

b)  $0.5 = a \times 10^{6.5 - \frac{130}{20}}$ [1]
    $0.5 = a \times 10^{0} = a$ [1]
    $a = 0.5$ [1]

c)  $d = 0.5 \times 10^{\left(6.5 - \frac{100}{20}\right)}$ [1]
    $d = 15.8\,\text{m}$ (3 s.f.) [1]

# Logarithms

## QUICK TEST (page 53)

1.

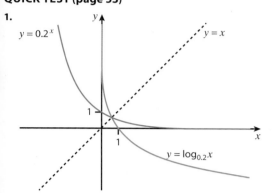

$y = 0.2^x$  $y = x$  $y = \log_{0.2} x$

2. a)  $q = \frac{1}{2}\ln 21 + \frac{1}{2}$
   b)  $q = \ln 21 e^{-1}$

3. a)  $\log_{2x} p = 4 \quad \rightarrow \quad p = 16x^4$
   b)  $\log 2 + \log 6 = \log 12$
   c)  $\frac{\ln 50 - \ln 2}{\ln e^2} = \ln 5$

## PRACTICE QUESTIONS (page 53)

1.  $\ln(x^2 + x - 2) - \ln(x + 2) - 3 = \ln\left(\frac{x^2 + x - 2}{x + 2}\right) - 3$ [1]
    $= \ln(x - 1) - 3$ [1]
    $\ln(x - 1) = 3$
    $x - 1 = e^3$ [1]
    $x = e^3 + 1$ [1]

2.  $Y = \log y \quad m = \log b \quad c = \log k$ [1]
    $m = \frac{\log 16 - \log 36p}{2} = \frac{1}{2}\log\left(\frac{16}{36p}\right) = \log\left(\frac{16}{36p}\right)^{\frac{1}{2}} = \log\left(\frac{2}{3\sqrt{p}}\right)$ [2]
    $\log 16 = \log\left(\frac{2}{3\sqrt{p}}\right) \times 4 + c$ [1]
    $= \log\left(\frac{16}{81p^2}\right) + c$ [1]
    $c = \log 16 - \log\left(\frac{16}{81p^2}\right) = \log 81p^2$ [1]
    $b = \frac{2}{3\sqrt{p}} \quad k = 81p^2$ [1]
    $y = 81p^2 \times \left(\frac{2}{3\sqrt{p}}\right)^x$ [1]

| $x$ | 2 | 4 | 6 | 8 |
|---|---|---|---|---|
| $y$ | $36p$ | 16 | $\frac{64}{9p^2}$ | $\frac{256}{81p^4}$ |

[1]

# Day 4

## Proof

### QUICK TEST (page 57)

1.  Proof by contradiction
    If rational then $\sqrt{3} = \frac{a}{b}$ where $a, b \in \mathbb{Z}$ and $a, b$ have no common factors

$3 = \frac{a^2}{b^2} \qquad a^2 = 3b^2$

$a^2$ divisible by 3

Therefore  $a$ divisible by 3

$a = 3k \qquad a^2 = 9k^2$

$9k^2 = 3b^2$

A contradiction, as there is a common factor.

2.  The starting assumption would be that both $a \geq \sqrt{c}$ and $b \geq \sqrt{c}$.

## PRACTICE QUESTIONS (page 57)

1.  Tallest blue-eyed person is Student F, who is 174.5 cm tall [1]
    Shortest non-blue-eyed person is also 174.5 cm (Student M) [1]
    Students M and F are of equal height, so Billie is incorrect that all the blue-eyed students are shorter than the non-blue-eyed students. [1]

2.  Make counter statement: Assume that $a^2 - 4b = 2$ [1]
    $a^2 = 2 + 4b = 2(1 + 2b)$ [1]
    $a^2$ is even, so $a$ must be an even number. Let $a = 2c$, where $c \in \mathbb{Z}$ [1]
    $a^2 - 4b = 2$
    $4c^2 - 4b = 2$
    $4(c^2 - b) = 2$ [1]
    $c^2$ is an integer and $b$ is an integer, so $c^2 - b$ must be an integer. [1]
    $c^2 - b = \frac{2}{4} = \frac{1}{2}$  This is not an integer $\therefore a^2 - 4b \neq 2$ [1]

# Differentiation

## QUICK TEST (page 63)

1. a)  $y = f''(x) = -3\sin x + e^x$
   b)  3
   c)  Convex, since $f''(x) \geq 0$ in this region

## PRACTICE QUESTIONS (page 63)

1. a)  Express each term in index form:
       $\left(x^{\frac{2}{3}} + 4e^{2x}\right)\left(x^{\frac{2}{3}} - 4e^{2x}\right)$ [1]
       Expansion: $x^{\frac{4}{3}} - 16e^{4x}$ [1 mark for each fully correct term]
   b)  $\frac{4}{3}x^{\frac{1}{3}} - 64e^{4x}$ [1 mark for each fully correct term]
   c)  Gradient $= \frac{4}{3}(8)^{\frac{1}{3}} - 64e^{4 \times 8}$
       Mark for clear attempt to substitute $x = 4$ into $\frac{dy}{dx}$ [1]
       Gradient $= \frac{8}{3} - 64e^{32}$ [1]

2. a)  $\frac{dy}{dx} = \frac{1}{10x} - 60x^4 + 3e^{3x}$ [1 mark for each fully correct term]
   b)  For substitution of $x = 1$ into derivative:
       $\frac{dy}{dx} = \frac{1}{10} - 60 + 3e^3$ [1]
       $a = 3e^3 - 59.9$ [1]
   c)  The gradient at $x = 1$ is $m_1 = 0.356610...$ [1]
       The gradient at $x = 1.1$ is
       $m_{1.1} = \frac{1}{11} - 60 \times (1.1)^4 + 3e^{3.3} = -6.417174...$ [1]
       Change of sign of the gradient function where the curve is continuous means that there has been a turning point. [1]
       As the gradient changes from positive to negative, the turning point is a maximum. [1]

# Differentiation and Combined Functions

## QUICK TEST (page 67)

1. a)  $\frac{dy}{dx} = 12(2x + 3)^5$
   b)  $\frac{dy}{dx} = \frac{3\cos x}{x} - \frac{3\sin x}{x^2}$

**c)** $\frac{dy}{dx} = 12xe^{4x} + 24x^2e^{4x}$

**d)** $\frac{dy}{dx} = (2\ln3)\,3^{2x} + 4\cos x - 4x\sin x$

**2. a)** $\frac{dy}{dx} = \frac{1}{y^2 - e^y - 2}$ \qquad gradient at $y = 2$ is $\frac{1}{2 - e^2}$

**b)** $\frac{dy}{dx} = \frac{1}{3 + \sec^2 y}$ \qquad gradient at $y = 2$ is $\frac{1}{3 + \sec^2 2}$

**3. a)** $\frac{dy}{dx} = -\frac{x}{4y}$ \quad at $(1, 2.5)$ $\frac{dy}{dx} = -\frac{1}{10}$ \quad at $(1, -2.5)$ $\frac{dy}{dx} = \frac{1}{10}$

**b)** $\frac{dy}{dx} = \frac{3}{2e^{2t}-1} = 1.5e^{1-2t}$ \quad at $(1, 1.5)$ $\frac{dy}{dx} = \frac{3}{2}$

**c)** $\frac{dy}{dx} = t + \cos(2t)$ \quad at $t = \frac{\pi}{3}$ coordinate $\left(\frac{2\pi - 12}{3}, \frac{\pi^2}{9} + \frac{\sqrt{3}}{2}\right)$

$\frac{dy}{dx} = \frac{2\pi - 3}{6} = \frac{\pi}{3} - \frac{1}{2}$

**d)** $\frac{dy}{dx} = \frac{6x}{e^y} = \frac{2}{x}$ \quad at $(1, \ln3)$ \qquad $\frac{dy}{dx} = 2$

**e)** $\frac{dy}{dx} = \frac{12\cos x - 2y}{2x}$ \quad at $\left(\frac{\pi}{2}, \frac{12}{\pi}\right)$ \qquad $\frac{dy}{dx} = \frac{-24}{\pi^2}$

**4. a)** $\frac{dy}{dx} = \frac{8t}{6e^{2t}}$

**b)** $(3, -3)$

## PRACTICE QUESTIONS (page 67)

**1.** For attempt to use chain rule: [1]
Let $y = \sin(u)$ where $u = e^{2x}$ [1]

$\frac{dy}{du} = \cos(u)$ \qquad $\frac{du}{dx} = 2e^{2x}$ [1]

$\frac{dy}{dx} = 2e^{2x}\cos(e^{2x})$ [1]

**2. a)** For clear attempt to use product rule:
$y = f(x)g(x)$ where $f(x) = 2x^2$ and $g(x) = \sin(e^{2x})$ [1]
$f'(x) = 4x$ \qquad $g'(x) = 2e^{2x}\cos(e^{2x})$ [1]

$\frac{dy}{dx} = 2x^2 \times 2e^{2x}\cos(e^{2x}) + 4x \times \sin(e^{2x})$ [1]

$= 4x^2e^{2x}\cos(e^{2x}) + 4x\sin(e^{2x})$ [1]

**b)** $\frac{dy}{dx} = 4(\ln\sqrt{\pi})^2\,(e^{2\ln\sqrt{\pi}})\cos(e^{2\ln\sqrt{\pi}}) + 4(\ln\sqrt{\pi})\sin(e^{2\ln\sqrt{\pi}})$ [1]

$e^{2\ln\sqrt{\pi}} = \pi$ [1]

$\sin(\pi) = 0$ and $\cos(\pi) = -1$ [1]

$\frac{dy}{dx} = -4\pi\,(\ln\sqrt{\pi})^2$ [1]

The function is decreasing at $x = \ln\sqrt{\pi}$ as $\frac{dy}{dx} < 0$ [1]

**3. a)** For expanding brackets (or using product rule):
$x^2y - xy^2 = 12$ [1]
For implicit differentiation of the two terms with use of product rule:
$\left(2xy + x^2\frac{dy}{dx}\right) - \left(y^2 + 2xy\frac{dy}{dx}\right) = 0$ [2]
For rearranging to get $\frac{dy}{dx}$
$x^2\frac{dy}{dx} - 2xy\frac{dy}{dx} = y^2 - 2xy$ [1]
$\frac{dy}{dx} = \frac{y^2 - 2xy}{x^2 - 2xy}$ [1]

**b)** For finding $y$-coordinate:
$4^2y - 4y^2 = 12$ [1]
$y^2 - 4y + 3 = 0$
$(y - 1)(y - 3) = 0$ [1]
$y = 1$ [1]
$y = 3$ [1]
At $(4, 1)$ \quad $\frac{dy}{dx} = \frac{1^2 - 2 \times 4 \times 1}{4^2 - 2 \times 4 \times 1} = -\frac{7}{8}$ [1]
At $(4, 3)$ \quad $\frac{dy}{dx} = \frac{3^2 - 2 \times 4 \times 3}{4^2 - 2 \times 4 \times 3} = \frac{-15}{-8} = \frac{15}{8}$ [1]

Equations of the normal:
Normal to curve at $(4, 1)$:
Has gradient $\frac{8}{7}$

$y - 1 = \frac{8}{7}(x - 4)$ \quad $\rightarrow$ \quad $y = \frac{8}{7}x - \frac{25}{7}$ [1]

Normal to curve at $(4, 3)$:
Has gradient $-\frac{8}{15}$

$y - 3 = -\frac{8}{15}(x - 4)$ \quad $\rightarrow$ \quad $y = -\frac{8}{15}x + \frac{77}{15}$ [1]

Solving simultaneous equations:
$\frac{8}{7}x - \frac{25}{7} = -\frac{8}{15}x + \frac{77}{15}$ [1]

$\frac{176}{105}x = \frac{914}{105}$ [1]

$x = \frac{457}{88}$ [1]

# Integration 1

## QUICK TEST (page 73)

**1. a)** 291.6 units$^2$ (1 d.p.)

**b)**

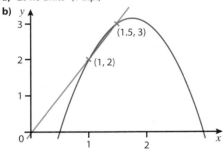

$\int_1^{1.5} y\,dx = \left[-\frac{2}{3}x^3 + \frac{7}{2}x^2 - 3x\right]_1^{1.5}$

Area of trapezium $= \frac{1}{2}(2 + 3) \times \frac{1}{2} = 1.25$

Bounded area $= \int_1^{1.5} y\,dx -$ area of trapezium

$= 1.291666\ldots - 1.25 = 0.0416666\ldots$

$= 0.417$ units$^2$ (3 s.f.)

**2.** $\frac{5}{2}e^{4x} - 15\sin\frac{x}{3} + \frac{9}{2}x^{\frac{2}{3}} + C$

## PRACTICE QUESTIONS (page 73)

**1. a)** $\int (a\sin2x + 2\cos ax)\,dx = -\frac{a}{2}\cos2x + \frac{2}{a}\sin ax + C$ [3]

**b)** $\frac{\sqrt{2}}{3} = -\frac{a}{2}\cos\left(2 \times \frac{\pi}{12}\right) + \frac{2}{a}\left(\sin\frac{a\pi}{12}\right) + C$ [1]

$\frac{\sqrt{2}}{3} = -\frac{a}{2} \times \frac{\sqrt{3}}{2} + \frac{2}{a}\sin\left(\frac{a\pi}{12}\right) + C$ [1]

$C = \frac{\sqrt{2}}{3} + \frac{a\sqrt{3}}{4} - \frac{2}{a}\sin\left(\frac{a\pi}{12}\right)$ [1]

**c)** $f\left(-\frac{\pi}{2}\right) = \frac{3a + 4}{6} + C = -\frac{a}{2}\cos(-\pi) + \frac{2}{a}\sin\left(-\frac{a\pi}{2}\right) + C$ [1]

$\frac{3a + 4}{6} = -\frac{a}{2}\cos(-\pi) + \frac{2}{a}\sin\left(-\frac{a\pi}{2}\right)$

$\frac{3a + 4}{6} = -\frac{a}{2} \times -1 + \frac{2}{a}\sin\left(-\frac{a\pi}{2}\right)$ [1]

If $a$ is an odd integer, the value of $\sin\left(-\frac{a\pi}{2}\right) = \pm 1$ [1]

$\frac{3a + 4}{6} = \frac{a}{2} \pm \frac{2}{a}$ [1]

$= \frac{a^2 \pm 4}{2a}$

$6a^2 + 8a = 6a^2 \pm 4 \times 6$ \qquad $a$ is positive so

$8a = 24$ \qquad $\rightarrow$ \qquad $a = 3$ [1]

$f'(0) - f(0) = (3\sin0 + 2\cos0) -$
$\left(-\frac{3}{2}\cos(0) + \frac{2}{3}\sin(0) + \frac{\sqrt{2}}{3} + \frac{3\sqrt{3}}{4} - \frac{2}{3}\sin\left(\frac{3\pi}{12}\right)\right)$ [2]

$= (2) - \left(-\frac{3}{2} + \frac{\sqrt{2}}{3} + \frac{3\sqrt{3}}{4} - \frac{\sqrt{2}}{3}\right)$

$= \left(\frac{14 - 3\sqrt{3}}{4}\right)$ [1]

**2. a)** $f'(y) = 4\sin\left(\frac{y}{2}\right) + 2\cos\left(\frac{y}{2}\right)$ [2]

**b)** $4\sin\left(\frac{y}{2}\right) + 2\cos\left(\frac{y}{2}\right) = -8\cos\left(\frac{y}{2}\right) + 4\sin\left(\frac{y}{2}\right) - 5$ **[1]**

$10\cos\left(\frac{y}{2}\right) = -5$

$\cos\left(\frac{y}{2}\right) = -\frac{1}{2}$ **[1]**

$\frac{y}{2} = \frac{2\pi}{3}$

$y = \frac{4\pi}{3}$ **[1]**

$x = 4\sin\left(\frac{2\pi}{3}\right) + 2\cos\left(\frac{2\pi}{3}\right) = -1 + 2\sqrt{3}$ **[1]**

$P = \left((-1 + 2\sqrt{3}), \frac{4\pi}{3}\right)$

**c)** $A = \int_{\frac{5\pi}{4}}^{\frac{4\pi}{3}} f'(y) - f(y)\, dy$ **[1]**

$= \int_{\frac{5\pi}{4}}^{\frac{4\pi}{3}} 4\sin\left(\frac{y}{2}\right) + 2\cos\frac{y}{2} - \left(-8\cos\left(\frac{y}{2}\right) + 4\sin\left(\frac{y}{2}\right) - 5\right) dy$

$= \int_{\frac{5\pi}{4}}^{\frac{4\pi}{3}} 10\cos\frac{y}{2} + 5\, dy$ **[1]**

$= \left[ 20\sin\frac{y}{2} + 5y \right]_{\frac{5\pi}{4}}^{\frac{4\pi}{3}}$ **[2]**

$= \left(20\sin\frac{2\pi}{3} + \frac{20\pi}{3}\right) - \left(20\sin\frac{5\pi}{8} + \frac{25\pi}{4}\right)$ **[2]**

$= \frac{20\sqrt{3}}{2} + \frac{20\pi}{3} - 20\sin\frac{5\pi}{8} - \frac{25\pi}{4}$ **[1]**

$= 10\sqrt{3} - 20\sin\frac{5\pi}{8} + \frac{5\pi}{12}$ **[1]**

# ay 5

## tegration 2

### JICK TEST (page 78)

**a)** $\frac{1}{3}e^{3x-2} + C$   **b)** $8\sin(7x-3) + C$

**c)** $\frac{1}{9}\ln(9x-1) + C$   **d)** $-\frac{1}{2}e^{1-2x} + C$

**a)** $\frac{3}{x+2}$   **b)** $\frac{3}{x} - \frac{3}{2x+4} - \frac{1}{2x-4}$

**a)** $2(2x-5)^{\frac{1}{2}} + C$   **b)** $\frac{1}{3}e^{\sin 3x} + C$

**c)** $\frac{3(\ln 2x)^2}{2} + C$

**a)** $y = \frac{x}{2} + \frac{1}{8}\sin(4x) + C$   **b)** $\frac{\pi - 1}{8}$

| | $u$ | $\frac{dv}{dx}$ |
|---|---|---|
| **a)** | $4x + 2$ | $\sin(2x)$ |
| **b)** | $2x$ | $e^x$ |
| **c)** | $x$ | $(x+3)^{\frac{1}{2}}$ |
| **d)** | $3x^2 + 2x + 1$ | $(x-3)^5$ |

| | $\frac{du}{dx}$ | $v$ |
|---|---|---|
| **a)** | $4$ | $-\frac{1}{2}\cos(2x)$ |
| **b)** | $2$ | $e^x$ |
| **c)** | $1$ | $\frac{2}{3}(x+3)^{\frac{3}{2}}$ |
| **d)** | $6x + 2$ | $\frac{1}{6}(x-3)^6$ |

**a)** $-(2x+1)\cos 2x + \sin 2x + C$

**b)** $2xe^x - 2e^x + C$

**c)** $\frac{2x}{3}(x+3)^{\frac{3}{2}} - \frac{4}{15}(x+3)^{\frac{5}{2}} + C$

**d)** $\frac{1}{6}(x-3)^6(3x^2 + 2x + 1) - \frac{1}{42}(x-3)^7(6x+2) + \frac{1}{56}(x-3)^8 + C$

**RACTICE QUESTIONS (page 79)**

$\int g(x)\, dx = \int \frac{2x(x+2)}{(x-1)(x^2+2x-3)}\, dx + \int 1\, dx$

Fully factorise the denominator: $(x+3)(x-1)^2$ **[1]**

Use partial fractions:

$\frac{2x^2 + 4x}{(x+3)(x-1)^2} = \frac{A}{x+3} + \frac{B}{x-1} + \frac{C}{(x-1)^2}$ **[1]**

$2x^2 + 4x = A(x-1)^2 + B(x+3)(x-1) + C(x+3)$

When $x = 1$:

$2 \times 1^2 + 4 \times 1 = C(1+3)$   →   $6 = 4C$   →   $C = \frac{3}{2}$

When $x = -3$:

$2 \times (-3)^2 + 4 \times (-3) = A((-3)-1)^2$   →   $6 = 16A$   →   $A = \frac{3}{8}$ **[1]**

When $x = 0$:

$2 \times (0)^2 + 4 \times (0) = \frac{3}{8}(-1)^2 + B(3)(-1) + \frac{3}{2}(3)$   →   $3B = \frac{39}{8}$   →   $B = \frac{13}{8}$ **[1]**

$\frac{2x(x+2)}{(x-1)(x^2+2x-3)} = \frac{3}{8(x+3)} + \frac{13}{8(x-1)} + \frac{3}{2(x-1)^2}$ **[1]**

$\int g(x)\, dx = \int \frac{3}{8(x+3)} + \frac{13}{8(x-1)} + \frac{3}{2(x-1)^2}\, dx + \int 1\, dx$

$= \frac{3}{8}\ln(x+3) + \frac{13}{8}\ln(x-1) - \frac{3}{2(x-1)} + x + C$ **[2]**

**2. a)** $\int f(\theta)\, d\theta = \int \theta^2 \cos\left(\theta + \frac{\pi}{4}\right) d\theta$

Use integration by parts:

Let $u = \theta^2$   →   $\frac{du}{d\theta} = 2\theta$

Let $\frac{dv}{d\theta} = \cos\left(\theta + \frac{\pi}{4}\right)$   →   $v = \sin\left(\theta + \frac{\pi}{4}\right)$ **[1]**

$\int \theta^2 \cos\left(\theta + \frac{\pi}{4}\right) d\theta = \theta^2 \sin\left(\theta + \frac{\pi}{4}\right) - \int 2\theta \sin\left(\theta + \frac{\pi}{4}\right) d\theta$ **[1]**

This requires a second application of integration by parts:

Let $u_2 = 2\theta$   →   $\frac{du_2}{dx} = 2$

Let $\frac{dv_2}{d\theta} = \sin\left(\theta + \frac{\pi}{4}\right)$   →   $v_2 = -\cos\left(\theta + \frac{\pi}{4}\right)$ **[1]**

$\int f(\theta)\, d\theta = \theta^2 \sin\left(\theta + \frac{\pi}{4}\right) - \int 2\theta \sin\left(\theta + \frac{\pi}{4}\right) d\theta = \theta^2 \sin\left(\theta + \frac{\pi}{4}\right)$

$-\left(-2\theta\cos\left(\theta + \frac{\pi}{4}\right) - \int -2\cos\left(\theta + \frac{\pi}{4}\right) d\theta\right)$

$= \theta^2 \sin\left(\theta + \frac{\pi}{4}\right) + 2\theta\cos\left(\theta + \frac{\pi}{4}\right) - 2\sin\left(\theta + \frac{\pi}{4}\right) + C$ **[3]**

**b)** $g(\theta) = \frac{d}{d\theta}\left(\frac{2}{27}\cos 3\theta + \frac{2\theta}{9}\sin 3\theta - \frac{\theta^2}{3}\cos 3\theta\right)$

$\frac{d}{d\theta}\left(\frac{2}{27}\cos 3\theta\right) = -\frac{2}{9}\sin 3\theta$ **[1]**

$\frac{d}{d\theta}\left(\frac{2\theta}{9}\sin 3\theta\right)$   Let $u = \frac{2\theta}{9}, \frac{du}{d\theta} = \frac{2}{9}$

Let $v = \sin 3\theta, \frac{dv}{d\theta} = 3\cos 3\theta$

$\frac{d}{d\theta}\left(\frac{2\theta}{9}\sin 3\theta\right) = \frac{2}{9}\sin 3\theta + \frac{2\theta}{3}\cos 3\theta$ **[2]**

$\frac{d}{d\theta}\left(-\frac{\theta^2}{3}\cos 3\theta\right)$   Let $u = -\frac{\theta^2}{3}, \frac{du}{d\theta} = \frac{2\theta}{3}$

Let $v = \cos 3\theta, \frac{dv}{d\theta} = -3\sin 3\theta$

$\frac{d}{d\theta}\left(-\frac{\theta^2}{3}\cos 3\theta\right) = -\frac{2\theta}{3}\cos 3\theta + \theta^2\sin 3\theta$ **[2]**

$g(\theta) = -\frac{2}{9}\sin 3\theta + \frac{2}{9}\sin 3\theta + \frac{2\theta}{3}\cos 3\theta - \frac{2\theta}{3}\cos 3\theta + \theta^2\sin 3\theta$

$= \theta^2 \sin 3\theta$ **[1]**

For each value of $\theta$ given there is only one intersection between the markings on the axis of the graph. So if the value of $\theta$ makes the equation $\theta^2\sin 3\theta = \theta^2\cos\left(\theta + \frac{\pi}{4}\right)$ hold true then the intersection is correct.

P: $\left(-\frac{5\pi}{8}\right)^2 \sin 3\left(-\frac{5\pi}{8}\right) = 1.475364878\ldots$

$\left(-\frac{5\pi}{8}\right)^2 \cos\left(\left(-\frac{5\pi}{8}\right) + \frac{\pi}{4}\right) = 1.475364878\ldots$

Q: $\left(-\frac{7\pi}{16}\right)^2 \sin\left(3\left(-\frac{7\pi}{16}\right)\right) = 1.57073254\ldots$

$\left(-\frac{7\pi}{16}\right)^2 \cos\left(\left(-\frac{7\pi}{16}\right) + \frac{\pi}{4}\right) = 1.57073254\ldots$

R: $(0)^2 \sin 3(0) = 0$

$(0)^2 \cos\left(0 + \frac{\pi}{4}\right) = 0$

As the equation is true for each value of $\theta$, the intersections are at $\theta_P = -\frac{5\pi}{8}, \theta_Q = -\frac{7\pi}{16}$ and $\theta_R = 0$ **[2]**

**c)** **A** – no. The graphs intersect so the area must be considered as two regions.

**B** – yes. Since the modulus is used, each part of the area is found independently, then the positive value of each is added to find the total area.

**C** – no. The two regions have been treated separately but one result will be positive and the other negative; by adding (without first taking the modulus) some of the relevant area will cancel out.

**D** – no.

**E** – yes.

**[2 marks if fully correct; 1 mark if at least three correct]**

3. First find the equation in Cartesian form:

$x^2 = 4\sin^2\theta$, $y^2 = \sin^2\theta\cos^{10}\theta$ **[1]**

$y^2 = \sin^2\theta\,(1-\sin^2\theta)^5$ **[1]**

$\sin^2\theta = \frac{x^2}{4}$

$y^2 = \frac{x^2}{4}\left(1-\frac{x^2}{4}\right)^5$ **[1]**

$y = \pm\frac{x}{2}\left(1-\frac{x^2}{4}\right)^{\frac{5}{2}}$ **[1]**

Find at least two roots ($x=-2$, $x=0$ and $x=2$) of the equation **[1]**
Use symmetry of the shape and find one of the quarters of the shaded area (any of the four is acceptable):

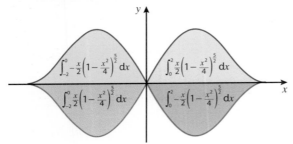

$\int_0^2 \frac{x}{2}\left(1-\frac{x^2}{4}\right)^{\frac{5}{2}} dx$

Integration by substitution or by recognising the form $\int f'(x)[f(x)]^n \, dx$:

Let $u = 1-\frac{x^2}{4}$ $\rightarrow$ $\frac{du}{dx} = -\frac{x}{2}$ $\rightarrow$ $dx = -\frac{2du}{x}$

$\int \frac{x}{2}\left(1-\frac{x^2}{4}\right)^{\frac{5}{2}} dx = \int \frac{x}{2}\left(u^{\frac{5}{2}}\right)\left(-\frac{2du}{x}\right)$

$= \int -\left(u^{\frac{5}{2}}\right) du$ **[2]**

New bounds of integration: $u = 1-\frac{0^2}{4} = 1$, and $u = 1-\frac{2^2}{4} = 0$ **[1]**

$\int_0^2 \frac{x}{2}\left(1-\frac{x^2}{4}\right)^{\frac{5}{2}} dx = \int_0^1 u^{\frac{5}{2}} du$

$= \left[\frac{2}{7}u^{\frac{7}{2}}\right]_0^1$

$= \left(\frac{2}{7}1^{\frac{7}{2}}\right) - \left(\frac{2}{7}0^{\frac{7}{2}}\right)$

$= \frac{2}{7}$ **[1]**

Area $= 4\times\frac{2}{7}$

$= \frac{8}{7}$ **[1]**

## Differential Equations

### QUICK TEST (page 83)

1. $\frac{dy}{dx} = 2y^4(x+3) \rightarrow \int y^{-4}\,dy = \int 2x+6\,dx \rightarrow -\frac{1}{3y^3} = x^2+6x+C$

$7x\frac{dy}{dx} - 5y^4 = 0 \rightarrow \int 7y^{-4}\,dy = \int 5x^{-1}\,dx \rightarrow -\frac{7}{3y^3} = 5\ln x + C$

---

$\frac{dx}{dy} = 2y^4(x+3) \rightarrow \int 2y^4\,dy = \int\frac{1}{x+3}\,dx \rightarrow \frac{2}{5}y^5 = \ln(x+3) + C$

$\frac{dx}{dy} - 4xy^2 = 2x\frac{dx}{dy} \rightarrow \int x^{-1} - 2\,dx = \int 4y^2\,dy \rightarrow \frac{4}{3}y^3 = \ln x - 2x + C$

$\sin x\frac{dx}{dy} = \frac{1}{xy^3\sin x} \rightarrow \int x\sin^2 x\,dx = \int y^{-3}\,dy \rightarrow$
$-\frac{1}{2y^2} = \frac{x^2}{4} - \frac{x}{4}\sin 2x + \frac{1}{8}\cos 2x + C$

$2x = \frac{\frac{dy}{dx}6y^2}{\sin x} \rightarrow \int 6y^2\,dy = \int 2x\sin x\,dx \rightarrow 2y^3 = -2x\cos x + 2\sin x + C$

2. **a)** $2y^4 + 3y = x^2 + 2x + 3$    **b)** $\sin\left(y-\frac{\pi}{8}\right) = \frac{1}{2}(x^2 - 3\sqrt{2})$

   **c)** $\frac{2}{5}y^5 = \ln(x+3) + \frac{2}{5}$    **d)** $\frac{7}{3y^3} = 2\ln x - \frac{331}{81}$

3. **a)** $\ln y = x^2 + 3x$

   Or equivalent: $y = e^{x^2+3x}$

   **b)** $\frac{e^{3y}}{3} = e^x + \frac{e^6}{3} - 1$ or equivalent

   **c)** $-\cos y = \ln(\sin x) - \frac{\sqrt{2}}{2}$ or equivalent

   **d)** $-\frac{1}{9}y^{-\frac{9}{2}} = x + 1$

   Or equivalent: $0 = x + \frac{y^{-\frac{9}{2}}}{9} + 1$

4. $-\frac{1}{2y^2} = \ln(4x^2 + x - 2) + C$

### PRACTICE QUESTIONS (page 83)

1. **a)** $\sin x\frac{dy}{dx} = (\cos x - 4\sin x)\tan y$

   $\frac{1}{\tan y}\frac{dy}{dx} = \frac{\cos x - 4\sin x}{\sin x}$

   $\cot y\frac{dy}{dx} = \cot x - 4$ **[1]**

   $\int\cot y\,dy = \int\cot x - 4\,dx$ **[1]**

   $\ln(\sin y) = \ln(\sin x) - 4x + C$

   **[3 marks: 1 for each integral and 1 for the constant of integration]**

   **b)** $\ln\left(\sin\frac{2\pi}{3}\right) = \ln\left(\sin\frac{2\pi}{3}\right) - 4\frac{2\pi}{3} + C$ **[1]**

   $C = \frac{8\pi}{3}$ **[1]**

   $\ln(\sin y) = \ln(\sin x) - 4x + \frac{8\pi}{3}$ **[1]**

2. $F = ma = m\frac{dv}{dt}$

   $\frac{1}{10}\frac{dv}{dt} = \frac{2\cos(2t-\pi)+2t}{3v^2}$    **[1 for use of $F = m\frac{dv}{dt}$]**

   $\frac{3}{10}v^2\frac{dv}{dt} = 2\cos(2t-\pi)+2t$    **[1 for separating variables]**

   $\int\left(\frac{3}{10}v^2\right)dv = \int(2\cos(2t-\pi)+2t)\,dt$

   $\frac{1}{10}v^3 = \sin(2t-\pi) + t^2 + C$

   **[3 marks for each integration and constant of integration]**

   At $(0, -2)$

   $\frac{1}{10}(-2)^3 = \sin(2(0)-\pi) + 0^2 + C$

   $C = -\frac{8}{10} = -\frac{4}{5}$ **[1]**

   $\frac{1}{10}v^3 = \sin(2t-\pi) + t^2 - \frac{4}{5}$

   At $t = 120$ **[1]**

   $\frac{1}{10}(v)^3 = \sin(2(120)-\pi) + 120^2 - \frac{4}{5}$ **[1]**

   $= 14398.25455...$

   $v^3 = 143982.5455...$ **[1]**

   $v = 52.41271004...$

   $= 52.4\text{ ms}^{-1}$ (3 s.f.) **[1]**

## Numerical Methods of Approximation

### QUICK TEST (page 91)

1. **a)** $f(0.5) = 3(0.5)^2 + 2\cos(0.5) - 3 = -0.49483...$
   $f(0.9) = 3(0.9)^2 + 2\cos(0.9) - 3 = 0.67321...$
   Change of sign of continuous function, so at least one root is within the interval.

**b)** $f(0.5) = \sin(2) + \cos^2\left(\frac{3}{4}\right) - 1 = 0.44466\ldots$

$f(0.6) = \sin(2.4) + \cos^2(1.08) - 1 = -0.10238\ldots$

Change of sign of continuous function, so at least one root is within the interval.

**c)** $f(-0.6) = \frac{6(-0.6)^2 - 2}{((-0.6) + 2)((-0.6) + 4)} = 0.033613\ldots$

$f(-0.5) = \frac{6(-0.5)^2 - 2}{((-0.5) + 2)((-0.5) + 4)} = -0.09523\ldots$

Change of sign of function over interval where the function is continuous (vertical asymptotes at $x = -2$ and $x = -4$), so at least one root is within the interval.

Any pair of valid rearrangements, e.g.

**a)** $x_{n+1} = \frac{1 - (x_n)^3}{3}$

$x_{n+1} = \sqrt[3]{1 - 3x}$

$x_{n+1} = \frac{1 - 3x}{x^2}$

**b)** $x_{n+1} = \frac{1}{4}\arcsin\left(-x_n^6\right)$

$x_{n+1} = \sqrt[6]{-\sin(4x_n)}$

$x_{n+1} = -\frac{\sin(4x_n)}{(x_n)^5}$

**a)** $2, 0.9882941\ldots, 0.55153252\ldots, 0.45166760\ldots, \ldots$

**b)** $-0.8, -0.23926442\ldots, -0.39106606\ldots, -0.3680703\ldots, \ldots$

**c)** $4, 2.9615, 2.2865, 1.9081, \ldots$

**a)** Correct, change of sign of function between $f(0.44795)$ and $f(0.44805)$ where function is continuous within interval.

**b)** Incorrect or impossible to tell, with justification for either:
No change of sign so no root or even number of roots. Justify 'incorrect' conclusion using a sketch of the graph or similar.

**c)** Incorrect as answer is given to 3 d.p. or 4 s.f.

**a)** Underestimate

**b)** $h = 1.5$

**c)** $y_1 = 4\ln 2$

$y_n = 4\ln 8 = 4\ln 2^3 = 12\ln 2$

## PRACTICE QUESTIONS (page 91)

**a)** $f(0) = \cos(0) - 0^{\frac{3}{4}} = 1$ **[1]**

$f\left(\frac{\pi}{3}\right) = \cos\left(\frac{\pi^2}{9}\right) - \left(\frac{\pi}{3}\right)^{\frac{3}{4}} = -0.578589\ldots$ **[1]**

The function is continuous and there is a change of sign, so there is a root within the interval $\left[0, \frac{\pi}{3}\right]$. **[1]**

Note: As a 'show that' question, clear working and a complete conclusion are required for full marks.

**b)** $x^{\frac{3}{4}} = \cos(x^2)$ **[1]**

$x = \left(\cos\left(x^2\right)\right)^{\frac{4}{3}}$

$p = \frac{4}{3}$ **[1]**

**c)**

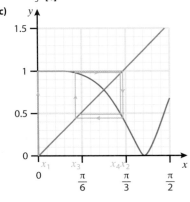

[3 marks if fully correct; 2 marks if three roots correct; 1 mark if one root correct]

**d)** $0, 1, 0.4400639\ldots, 0.9751541\ldots, 0.4847281\ldots, 0.9635332\ldots$ **[3]**

**e)** For any valid method. Iteration with this formula is very slow, so best avoided. Other options are to find a different iterative formula or to try Newton-Raphson as shown below:

$f(x) = \cos(x^2) - x^{\frac{3}{4}}$

$f'(x) = -2x\sin(x^2) - \frac{3}{4}x^{-\frac{1}{4}}$ **[1]**

$x_{n+1} = x_n - \frac{\cos\left(x_n^2\right) - x_n^{\frac{3}{4}}}{-2x_n\sin\left(x_n^2\right) - \frac{3}{4}x_n^{-\frac{1}{4}}}$ **[1]**

Let $x_1 = \frac{\pi}{3}$ **[1 mark for any value in the range $(0, \frac{\pi}{3}]$ with follow-through marks for subsequent values]**

$x_2 = 0.8250665\ldots$

$x_3 = 0.7765375\ldots$

$x_4 = 0.7744005\ldots$

$x_5 = 0.7743966\ldots$ **[3]**

Check accuracy of interval for 0.7744 as the root to 4 d.p.

$f(0.77435) = 0.00007809\ldots = $ positive

$f(0.77445) = -0.00008926\ldots = $ negative **[1]**

Change of sign of continuous function, so root is 0.7744 (4 d.p.) **[1]**

**2. a)** $h = \frac{60 - 30}{5 - 1} = 7.5$ **[1]**

| | $x_0$ | $x_1$ | $x_2$ | $x_3$ | $x_4$ |
|---|---|---|---|---|---|
| | 30 | 37.5 | 45 | 52.5 | 60 |
| | $8.98038514\ldots$ | $10.41943422\ldots$ | $12.2632371\ldots$ | $14.7633114\ldots$ | $18.4032290\ldots$ |
| | $y_0$ | $y_1$ | $y_2$ | $y_3$ | $y_4$ |

(For each mistake in table (or set of results) lose 1 mark. If no individual results shown, any mistake loses all marks automatically.) **[5]**

$\int_{30}^{60} \tan\left(x^{\frac{1}{10}}\right) + 3\,dx \approx \frac{1}{2} \times h \times \{(y_0 + y_n) + 2(y_1 + y_2 + \ldots + y_{n-1})\}$

$\approx 383.53342\ldots$ **[1]**

**b)** As the curve is convex in the interval [30, 60], the initial estimate will be an overestimate. **[1]**

By increasing the number of ordinates, a better approximation will be obtained and its value will be lower than that found in part **a)**. **[1]**

## Modelling, Quantities and Units

### QUICK TEST (page 93)

**1. a)** 189 120 seconds; any reasonable example about length of **time**

**b)** 4500 kg; any reasonable example about **mass** (NB: not weight)

**c)** 0.36 Nm; any reasonable example about **moments** or **turning force**

**d)** $19.4\,\text{ms}^{-1}$ (3 s.f.); any reasonable example about **velocity**

**e)** 72 N; any reasonable example about **force**

# Day 6

## Vectors

### QUICK TEST (page 99)

**1. a)** $\mathbf{p} = -2\mathbf{i} + \frac{3}{4}\mathbf{j} + \frac{\sqrt{2}}{2}\mathbf{k}$

**b)**

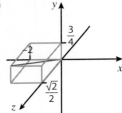

**c)** $|\mathbf{p}| = \sqrt{(-2)^2 + \left(\frac{3}{4}\right)^2 + \left(\frac{\sqrt{2}}{2}\right)^2}$

$= \sqrt{\frac{81}{16}} = 2.25$

**d)** $\cos\alpha = -\frac{2}{2.25}$

$\alpha = 152.7°$ (1 d.p.)

**2.** $\begin{pmatrix} 8 \\ -5 \\ 3 \end{pmatrix}$

**3. a)** $\overrightarrow{BA} = 15\mathbf{i} - 14\mathbf{j} + 2\mathbf{k}$   **b)** $\overrightarrow{AB} = -15\mathbf{i} + 14\mathbf{j} - 2\mathbf{k}$

## PRACTICE QUESTIONS (page 99)

**1. a)** Distance travelled BC is

$\sqrt{120^2 + 480^2 + (-160)^2} = \sqrt{270400} = 520\,\text{m}$

So distance AB is $620 - 520 = 100$ **[1]**

$(2a)^2 + (a - 110)^2 = 100^2$ **[1]**

$4a^2 + a^2 - 220a + 12100 = 10000$

$5a^2 - 220a + 2100 = 0$ **[1]**

$a = \frac{220 \pm \sqrt{(-220)^2 - 4 \times 5 \times 2100}}{10} = 22 \pm \frac{\sqrt{6400}}{10} = 30\text{ or }14$ **[1]**

$a = 30$ (as $14 < 20$) **[1]**

$\overrightarrow{AC} = (2a + 120)\mathbf{i} + (480)\mathbf{j} + (a - 270)\mathbf{k}\,\text{m}$ **[1]**

$\overrightarrow{AC} = 180\mathbf{i} + 480\mathbf{j} - 240\mathbf{k}\,\text{m}$ **[1]**

**b)** $\overrightarrow{AC} = 180\mathbf{i} + 480\mathbf{j} - 240\mathbf{k}\,\text{m}$

So distance AC is $\sqrt{180^2 + 480^2 + (-240)^2}$ **[1]**

$= 60\sqrt{89}\,\text{m}$ **[1]**

**c)** Collinear means $\overrightarrow{AD} = p(\overrightarrow{AC})$, where $p$ is a constant. **[1]**

Using the ratio $\pm 20\overrightarrow{AC} = \overrightarrow{AD}$ $\qquad \overrightarrow{AD} = \pm\frac{1}{20}\overrightarrow{AC}$ **[1]**

$\overrightarrow{AD} = \pm\frac{1}{20}(180\mathbf{i} + 480\mathbf{j} - 240\mathbf{k})$ **[1]**

$p = -\frac{1}{20}$ as the submarine surfaces and so $\overrightarrow{AD}$ must have a positive $\mathbf{k}$ component. **[1]**

$\overrightarrow{AD} = -9\mathbf{i} - 24\mathbf{j} + 12\mathbf{k}\,\text{m}$ **[1]**

**d)** $21 \times 12$   (or $12\,\text{m} + 240\,\text{m}$) **[1]**

$252\,\text{m}$ **[1]**

## Kinematics

### QUICK TEST (page 105)

**1. a)** $q = -1,\ p = -2$

**b)** $t = 4\,\text{s}$

**2. a)** $v = (-0.2t\sin(0.1t^2))\mathbf{i} + (2 + 2\ln t)$

**b)** $a = (-0.2\sin(0.1t^2) - 0.04t^2\cos(0.1t^2))\mathbf{i} + \left(\frac{2}{t}\right)\mathbf{j}$

**c)** $a = -0.225\mathbf{i} + 1\mathbf{j}\,\text{ms}^{-2}$ (3 s.f.)

**d)** $|a| = 1.02\,\text{ms}^{-2}$ (3 s.f.)

## PRACTICE QUESTIONS (page 105)

**1. a)** $\mathbf{r} = (3t^2 - 2t)\mathbf{i} + \left(\frac{t}{10}\ln t^2\right)\mathbf{j}$

$v = \left(\frac{d}{dt}(3t^2 - 2t)\right)\mathbf{i} + \left(\frac{d}{dt}\left(\frac{t}{10}\ln t^2\right)\right)\mathbf{j}$ **[1]**

$\frac{d}{dt}\left(\frac{t}{10}\ln t^2\right) = \frac{d}{dt}\left(\frac{t}{5}\ln t\right)$

Product rule:

Let $u = \frac{t}{5}$   $\qquad \frac{du}{dt} = \frac{1}{5}$

$v = \ln t$   $\qquad \frac{dv}{dt} = \frac{1}{t}$ **[1]**

$\frac{dy}{dt} = u\frac{dv}{dt} + v\frac{du}{dt} = \frac{t}{5} \times \frac{1}{t} + \frac{1}{5} \times \ln t = \frac{1}{5} + \frac{\ln t}{5}$ **[1]**

$v = (6t - 2)\mathbf{i} + \left(\frac{1}{5} + \frac{\ln t}{5}\right)\mathbf{j}$ **[1]**

$\mathbf{a} = \left(\frac{d}{dt}(6t - 2)\right)\mathbf{i} + \left(\frac{d}{dt}\left(\frac{1}{5} + \frac{\ln t}{5}\right)\right)\mathbf{j}$ **[1]**

$\mathbf{a} = 6\mathbf{i} + \frac{1}{5t}\mathbf{j}\,\text{ms}^{-2}$ **[1]**

**b)** At $t = 4\,\text{s}$

$v = (6 \times 4 - 2)\mathbf{i} + \left(\frac{1}{5} + \frac{\ln 4}{5}\right)\mathbf{j} = 22\mathbf{i} + 0.4772588...\mathbf{j}$ **[1]**

$|v| = \sqrt{22^2 + (0.4772588...)^2}$ **[1]**

$= 22.00517612... = 22.0\,\text{ms}^{-1}$ (3 s.f.) **[1]**

**c)** Position vector horizontally $3t^2 - 2t$

So $2 < 3t^2 - 2t < 4$ and $t \geqslant 0$ **[1]**

$3t^2 - 2t < 4$

$3t^2 - 2t - 4 < 0 \Rightarrow \frac{1 - \sqrt{13}}{3} < t < \frac{1 + \sqrt{13}}{3}$ (1.535...) **[1]**

$2 < 3t^2 - 2t$

$3t^2 - 2t - 2 > 0 \Rightarrow t > \frac{1 + \sqrt{7}}{3}$ (1.215...) or $t < \frac{1 - \sqrt{7}}{3}$ **[1]**

So within 1m of $x = 3$ for $1.535 - 1.215$ **[1]** $= 0.32\,\text{s}$ **[1]**

An alternative solution is possible by solving $\left|3t^2 - 2t - 3\right| < 1$.

## Mechanics

### QUICK TEST (page 111)

**1. a)** $3.79\,\text{N}$, at an angle of $69.7°$ below the positive $x$-direction

**b)** $\frac{16\sqrt{5}}{5}\,\text{N} = 7.16\,\text{N}$ (3 s.f.), at angle $\arctan 2 = 63.4°$ below the positive $x$-direction

**2. a)** $0.019\,\text{ms}^{-2}$ (3 d.p.)   **b)** $\frac{16}{5} = 3.2\,\text{ms}^{-2}$

### PRACTICE QUESTIONS (page 111)

**1. a)** Force diagram:

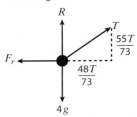

For resolving force   **[1]**

$F = ma$

Vertically:

$R + \frac{55T}{73} = 4g$ **[1]**

$R = 4g - \frac{55T}{73}$

Horizontally:

$F = ma$

$\frac{48T}{73} - F_r = 4 \times 0.5$ **[1]**

$\frac{48T}{73} - 2 = \mu\left(4g - \frac{55T}{73}\right)$ **[1]**

$\mu = \frac{\frac{48T}{73} - 2}{4g - \frac{55T}{73}} = \frac{48T - 146}{292g - 55T}$ **[1]**

**b)** $0 \le \mu \le 1$

$0 = \frac{48T - 146}{292g - 55T}$ $\rightarrow$ $48T = 146$ **[1]**

$T = 3.04\ldots$

$1 = \frac{48T - 146}{292g - 55T}$ $\rightarrow$ $48T - 146 = 292 \times 9.8 - 55T$ **[1]**

$103T = 3007.6$ **[1]**

$T = 29.2$

$3.04\ldots \le T \le 29.2$ **[1]**

**c)**

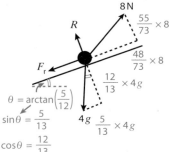

$\theta = \arctan\left(\frac{5}{12}\right)$

$\sin\theta = \frac{5}{13}$

$\cos\theta = \frac{12}{13}$

$\mu = \frac{48T - 146}{292g - 55T} = \frac{595}{3027} = 0.0982\ldots$ **[1]**

Perpendicular to slope; forces are balanced:

$R + \frac{55}{73} \times 8 = \frac{12}{13} \times 4 \times 9.81$ **[1]**

$R = 30.19414\ldots$ N **[1]**

Parallel to slope; forces are balanced:

$F = \frac{48}{73} \times 8 - \frac{5}{13} \times 4 \times 9.8 - \mu R$ **[1]**

$= -12.780\ldots$

$4a = -12.780\ldots$ **[1]**

$a = -3.195\ldots \text{ms}^{-2}$ **[1]**

$s = ?, u = 8, v = 0, a = -3.195\ldots$

$v^2 = u^2 + 2as$

$0 = 8^2 + s \times 2 \times 3.195\ldots$ **[1]**

$s = \frac{64}{2 \times 3.195\ldots} = 10.015 = 10.0 \text{ m (2 s.f.)}$ **[1]**

## Moments

### QUICK TEST (page 117)

**a)** 0.75 Nm (anticlockwise)

**b)** 9.8 Nm (clockwise)

**c)** $\frac{5\sqrt{2}}{2}$ Nm = 3.54 Nm (3 s.f.) (anticlockwise)

**d)** $(-30 + 30\sqrt{2})$ Nm = 12.4 Nm (3 s.f.) (clockwise)

**a)** $x = 1.5$ m   **b)** $\theta = 30°$   **c)** $P = -7\sqrt{2}$ N

**a)** 8 Nm (clockwise)   **b)** 0 Nm

**c)** 2 Nm (clockwise)

**a)** $x = 2m$, $R_A = 10$ N

**b)** $T = 3$ N, $x = \frac{160}{7}$ cm = 22.9 cm (3 s.f.)

### PRACTICE QUESTIONS (page 117)

**a)**

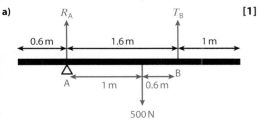

M(A)      $P\circlearrowleft = P\circlearrowright$

$1 \times 500 = 1.6 \times T_B$

(moments about B also acceptable) **[1]**

$T_B = \frac{500}{1.6} = 312.5$ N **[1]**

Vertically $R_A + T_B = 500$ N

$R_A = 500 - 312.5 = 187.5$ N **[1]**

**b)** The greatest load it can safely take is when the load is at one end of the bridge.

Test at end near A: M(A)        $P\circlearrowleft = P\circlearrowright$

$1 \times 500 = 0.6 \times F$

$F = 833.3333$ N **[2 marks: 1 for justifying that the load at end near A would be larger than that at end near B; it doesn't have to be fully calculated. 1 mark for answer]**

Test at end near B: M(B)   $P\circlearrowleft = P\circlearrowright$

$1 \times F = 0.6 \times 500$

$F = 300$ N **[2]**

The maximum load the bridge could safely take is 300 N. **[1]**

**c)** By considering symmetry, supports will be equal distance from their respective ends of the plank (can be implied) **[1]**

M(B)  $P\circlearrowleft = P\circlearrowright$

$500(1.6 - x) = 150g \times x$ **[1]**

$800 - 500x = 1470x$ **[1]**

$1970x = 800$

$x = \frac{80}{197} = 0.40609\ldots$ **[1]**

$AB = 3.2 - 2x = 2.4$ m (2 s.f.) **[1]**

**2.** Add forces to diagram:

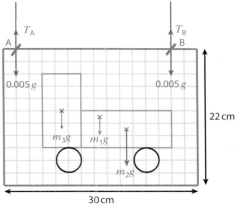

Conversion: 300 g m$^{-2}$ = 0.03 g cm$^{-2}$ = 0.00003 kg cm$^{-2}$ **[1]**

$m_1 = 30 \times 22 \times 0.00003 = 0.0198$ kg

$m_2 = 14 \times 6 \times 0.00003 = 0.00252$ kg

$m_3 = 12 \times 6 \times 0.00003 = 0.00216$ kg **[2]**

Take moments about A or B:     M(A)        $P\circlearrowleft = P\circlearrowright$

$0.07 \times m_3 g + 0.13 \times m_1 g + 0.17 \times m_2 g + 0.24 \times 0.005g = 0.24 \times T_B$ **[2]**

$T_B = \frac{0.042665\ldots}{0.24} = 0.177772\ldots$ N **[1]**

$= 0.178$ N (3 s.f.) **[1]**

Moments or vertical equilibrium: M(B)        $P\circlearrowleft = P\circlearrowright$

$0.24 \times T_A = 0.07 \times m_2 g + 0.11 \times m_1 g + 0.17 \times m_3 g + 0.24 \times 0.005g$ **[1]**

$T_A = \frac{0.03843\ldots}{0.24} = 0.1601\ldots$ N **[1]**

$= 0.160$ N (3 s.f.) **[1]**

# Day 7

## Correlation

### QUICK TEST (page 119)

1. $f = -6.8 - 68g$
2. $a = 2.24, b = 0.759$
3. a) Strong positive correlation
   b) Weak negative correlation
4. $r = 0.638$
5. $H_0 : \rho = 0, H_1 : \rho < 0$
   $r = -0.959$
   Reject $H_0$, $-0.959 < -0.7293$, evidence at 5% level to suggest $p$ and $q$ are negatively correlated
6. Accept $H_0$, $0.5 < 0.5155$, no evidence at 1% level to suggest correlation

### PRACTICE QUESTIONS (page 119)

1. a) Take logs of both sides for $y = ax^b$ to give $\log y = \log a + b \log x$
   Compare coefficients:

   $\log y = 0.78 + 2.62 \log x,$    $0.78 = \log a \to a = 10^{0.78} \to a = 6.03$

   $\log y = \log a + b \log x$    $b = 2.62$ **[3]**

   $y = 6.03x^{2.62}$ **[1]**

   b) Substitute $x = 5$ into $y = 6.03x^{2.62}$ to give $y = 408.9$ **[1]**
   Therefore 40 890 bacteria **[1]**

2. a) The PMCC is a value between $-1$ and $+1$, which measures the strength of the correlation. **[1]**
   b) From calculator, $r = 0.791$ **[1]**
   c) $H_0 : \rho = 0, H_1 : \rho \neq 0$ (two-tailed test) **[2]**
   Critical value from the table is 0.6664 for $n = 9$ **[1]**
   Reject $H_0$ as $0.791 > 0.6664$ and conclude there is evidence to suggest the weight of the grape crop and the number of hours of sunshine a vineyard had are correlated. **[2]**

## The Normal Distribution

### QUICK TEST (page 123)

1. a) 0.0228   b) 0.0228   c) 0.4207   d) 0.7881
2. a) 0.2902   b) 0.0878
3. a) 629.047   b) 618.518   c) 628.587
4. $\mu = 39.8$
5. $\sigma = 14.8$
6. $\mu = 101.4$
7. $\sigma = 7.63$
8. $\mu = 206.3, \sigma = 20.5$

### PRACTICE QUESTIONS (page 123)

1. $X \sim N(150, 30)$
   a) Using calculator $P(130 < X < 170)$ **[1]**
      $= 0.495$ **[1]**
   b) $P(X > x) = 0.06$, therefore $P(X < x) = 0.94$ **[1]**
      Using calculator, $x = 197$ g (3 s.f.) **[1]**
   c) Using calculator, $P(X > 140) = 0.6305...$ **[1]**
      $0.63055... \times 5000 = 3152.79...$ **[1]**
      3153 apples will weigh more than 140 g. **[1]**
2. $X \sim N(50, \sigma^2)$
   From question    $P(48 < X < 52) = 1 - 0.015$ **[1]**

See diagram    $P(X < 52) = 0.5 + \left(\frac{0.985}{2}\right) = 0.9925$ **[1]**

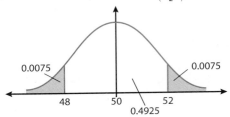

Using calculator    $Z = 2.432...$ **[1]**

Using $Z = \frac{X - \mu}{\sigma}$    $2.432 = \frac{52 - 50}{\sigma}$ **[1]**

$\sigma = 0.822$ (3 s.f.) **[1]**

3. $X \sim N(\mu, \sigma^2)$, $P(X < 25\,000) = 0.1$, and $P(X > 50\,000) = 0.05$

Initial diagram:

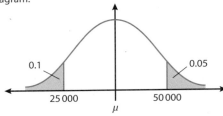

$Z$ values found from calculator or percentage points table, added to diagram:

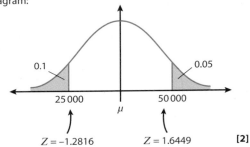

$Z = -1.2816$    $Z = 1.6449$ **[2]**

Simultaneous equations formed:
$$-1.2816\sigma = 25\,000 - \mu$$
$$1.6449\sigma = 50\,000 - \mu \quad \textbf{[2]}$$

Simultaneous equations solved to give:
$$\mu = 35\,948.2$$

$$\sigma = 8542.6 \text{ (1 d.p.)} \quad \textbf{[1]}$$

## The Normal Approximation and Hypothesis Testing

### QUICK TEST (page 127)

1. 0.9296
2. 0.987
3. 0.08483
4. 0.03798
5. Reject $H_0$ as $0.0786 < 0.1$ and conclude there is evidence to suggest that $\mu > 50$
6. Accept $H_0$ as $0.0368 > 0.01$ and conclude there is no evidence to suggest that $\mu < 85$
7. Reject $H_0$ as $0.0127 < 0.025$ and conclude there is evidence to suggest that $\mu \neq 200$

$\frac{62-60}{\sqrt{1.5}} = 1.633$

Accept $H_0$ as $1.633 < 1.6449$ and conclude there is no evidence to suggest that $\mu > 60$

## ACTICE QUESTIONS (page 127)

$X \sim N(240, 0.5^2)$

**a)** $P(239 < X < 241) = 0.9545$ are not unsuitable **[1]**

$1 - 0.9545 = 0.0455$, probability of being unsuitable **[1]**

**b)** Now this is a binomial distribution $Y \sim B(10, 0.0455)$ **[1]**

$P(Y = 0) = 0.6277$ (from binomial using calculator) **[1]**

**c)** Another binomial distribution $Z \sim B(100, 0.6277)$, find $P(Z \geqslant 70)$ **[1]**

Approximate by a normal distribution $L \sim N(62.77, 4.8341^2)$ **[1]**

Find $P(L > 69.5) = 0.08194$ **[2]**

$H_0 : \mu = 45, H_1 : \mu \neq 45$ **[1]**

$\bar{X} \sim N\left(45, \frac{4.4}{30}\right)$ **[1]**

**Either:**

$z = \frac{44 - 45}{\sqrt{\frac{4.4}{30}}} = -2.611$

2.5% critical value is $-1.96$ (two-tailed)

Reject $H_0$ as $-2.611 < -1.96$ and conclude there is evidence to suggest the mean lifetime of a battery has changed.

**Or:**

Find, using the normal distribution calculator function $P(X < 44) = 0.00451$ (3 s.f.)

Reject $H_0$ as $0.00451 < 0.05$ and conclude there is evidence to suggest the mean lifetime of a battery has changed.

Either method **[3]**

## robability

### JICK TEST (page 129)

**a)** $\frac{1}{10}$        **b)** $\frac{3}{10}$

**a)** $\frac{4}{9}$   **b)** $\frac{17}{90}$   **c)** $\frac{67}{90}$

**a)** 0.15        **b)** $\frac{3}{7}$

$\frac{9}{21}$ is the relative frequency. 21 is not a large enough sample to act as an estimate. Rain has been assumed to be independent and may not be.

$\frac{1}{3}$

0.2

**1. a)** $P(R|Q) = \frac{P(R \cap Q)}{P(Q)} \rightarrow \frac{4}{7} = \frac{\frac{1}{3}}{q} \rightarrow q = \frac{7}{12}$ **[2]**

**b)**

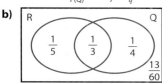

$P(Q|R') = \frac{\frac{1}{4}}{\frac{7}{15}} = \frac{15}{28}$ **[3]**

**c)** $P(R) \times P(Q) = \frac{8}{15} \times \frac{7}{12} = \frac{14}{45}$

$P(R \cap Q) = \frac{1}{3}$ **[2]**

Therefore R and Q are not independent since

$P(R) \times P(Q) \neq P(R \cap Q)$ **[1]**

**2. a)**

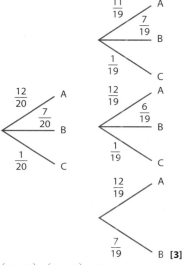

**[3]**

**b)** $\left(\frac{12}{20} \times \frac{11}{19}\right) + \left(\frac{7}{20} \times \frac{6}{19}\right) = \frac{87}{190}$ **[2]**

**c)** $P(C_2 | B_1) = \frac{P(C_2 \cap B_1)}{P(B_1)} = \frac{\frac{7}{20} \times \frac{1}{19}}{\frac{7}{20}} = \frac{1}{19}$, or directly from tree diagram **[2]**

# Index

**Acknowledgements**

The authors and publisher are grateful to the copyright holders for permission to use quoted materials and images. Every effort has been made to trace copyright holders and obtain their permission for the use of copyright material. The authors and publisher will gladly receive information enabling them to rectify any error or omission in subsequent editions. All facts are correct at time of going to press.

Published by Letts Educational
An imprint of HarperCollinsPublishers
1 London Bridge Street
London SE1 9GF

© HarperCollinsPublishers Limited 2019

ISBN 9780008276089

First published 2019

10 9 8 7 6 5 4 3 2 1

British Library Cataloguing in Publication Data.

A CIP record of this book is available from the British Library.

Authors: Rosie Benton and Sharon Faulkner
Commissioning Editor: Gillian Bowman
Project Management and Editorial: Richard Toms
Consultant: Deborah Dobson
Indexer: Simon Yapp
Inside Concept Design: Ian Wrigley
Cover Design: Sarah Duxbury
Text Layout: QBS Learning
Production: Karen Nulty
Printed and bound in Italy by Grafica Veneta